THE GREAT OIL
CONSPIRACY

THE GREAT OIL CONSPIRACY

*How the US Government Hid
the Nazi Discovery of Abiotic Oil
from the American People*

JEROME R. CORSI, PH.D.

Skyhorse Publishing

Skyhorse Publishing books may be purchased in bulk at special discounts
for sales promotion, corporate gifts, fund-raising, or educational purposes.
Special editions can also be created to specifications. For details, contact the
Special Sales Department, Skyhorse Publishing,
307 West 36th Street, 11th Floor, New York, NY 10018 or
info@skyhorsepublishing.com.

Skyhorse® and Skyhorse Publishing® are registered trademarks of Skyhorse
Publishing, Inc.®, a Delaware corporation.

Visit our website at www.skyhorsepublishing.com.

10 9 8 7 6 5 4 3 2 1

Library of Congress Cataloging-in-Publication Data is available on file.

ISBN: 978-1-62087-162-1

Printed in the United States of America

TABLE OF CONTENTS

Introduction

When I grew up in the 1950s during Dwight Eisenhower's presidency, my father bought a two-door Plymouth coupe for the family car.

Even back then we were told the United States was running out of oil, having supplied the Allies in World War II with the fuel needed to fight a war on two fronts and defeat simultaneously the Nazis in Germany and Imperial Japan.

But it didn't make sense to me: why did President Eisenhower want to build an interstate freeway system if we were running out of oil? Clearly, the government and the major oil companies had to know something we didn't.

Then I remember reading in a science magazine at the public library that the Russians had found oil thousands of meters below surface of the earth.

How did all the dinosaurs get that deep within the earth? Besides, how many dinosaurs exactly did it take to make a barrel of oil? For these questions, I found no satisfactory answers.

The purpose of this book is to expose to readers the Nazi secret of synthetic oil and the suppressed truth that oil is abiotic: not organic in nature.

It's a myth that hydrocarbon fuels are scarce. The truth is that proven reserves of oil and natural gas worldwide are greater today than ever in human history, despite increased demand from emerging economies in countries such as Brazil, Russia, India, and China – known together under the acronym of "BRIC" countries.

Moreover, non-traditional oil production is making great strides as the United States learns to make oil from the nation's abundant shale supply and offshore oil exploration; production has never been more robust. Off every major continent today, oil and natural gas are being discovered at deep-water and deep-earth levels.

Though most Americans have been indoctrinated by the politically correct media to believe we have nearly depleted our continental resources of oil and natural gas, the truth is that the United States is on its way to energy independence and could take major strides in the next few years towards surpassing Saudia Arabia and Russia as the world's leading oil and natural gas producer.

The scope of this book will not permit a thorough debunking of two other politically correct myths advanced by the enemies of hydrocarbon fuels, but with co-author Craig Smith, we tackled these subjects when collaborating in 2006. In *Black Gold Stranglehold: The Myth of Scarcity and the Politics of Oil*, it was proven that:

- There is no definitive proof that global warming is occurring, or that human activity in consuming hydrocarbon fuels contributes to any statistically significant "greenhouse gas" effect; and

- There is no definitive proof that consuming hydrocarbon fuels is inherently detrimental to the environment, as long as adequate precautions are taken in developing and producing energy resources and a determination is made by industry to develop and exploit "clean energy," including clean coal.

What this book explains is that Nazi scientists understood the fundamental chemical equations that explain how hydrocarbon fuels are produced without the assistance of any dead and decomposing living organism. And evidence shows that the United States still today has available hydrocarbon fuel resources, in both traditional and non-conventional reserves now being accessed through technological advances. The country has the ability not only to be energy independent but also to once again be the world's leading producer of oil and natural gas.

Breaking the regulatory grasp that government has created over decades and encouraging independent energy industry innovation and entrepreneurship are critical if energy prices in the future are going to remain affordable, so that the US economy can resume robust growth.

The United States government and major oil companies have perpetuated this fraud, encouraging the American people to believe that oil and natural gas are "fossil fuels" that will soon be depleted worldwide. But the truth is that hydrocarbon fuels, properly understood, are renewable fuels naturally produced by the earth on a continuing and abundant basis.

1

The Nazi Secret Science of Synthetic Oil

As the Allied armies raced to Berlin, and World War II drew to a close, the US Army had more than 3,000 separate teams involving 10,000 investigators (including industrialists, engineers, scientists, and technicians) visiting thousands of enemy factories, scientific institutions, businesses, and factories to conduct top-secret interviews and cart away trunk loads of captured documents.

"By the last month of the fighting in Germany, as the Allied armies rolled across the Rhine, combat-weary GIs were used to seeing groups of intelligence officers moving about the war zone," wrote professor of history Arnold Krammer.[1]

---

[1] Arnold Krammer, Professor of History at Texas A&M University, "Technology Transfer as War Booty: The US Technical Oil Mission to Europe, 1945," The Society for the History of Technology, 1981, published in Technology and Culture, Vol. 22, No. 1 (January 1981), pp 68-103, also available at http://www.jstor.org/discover/10.2307/3104293?uid=3739808&uid=2129&uid=2&uid=70&uid=4&uid=3739256&sid=47698781695217.

"They were no longer startled to see small groups of scholarly looking American officers drive up to bombed-out and newly captured factories and, apparently unmindful of the smoke and sometimes nearby gunfire, systematically investigate the plant."

The GIs watched, Krammer noted, as tons of records were "hauled out into the open for eventual crating and shipment" as German scientists were questioned by "soldiers" who wore neither rank nor unit designations on their American uniforms. The investigators were intelligence operatives – industrial scientists and government experts – and the German scientists they sought out had one thing in common – they had produced strategic materials for the Third Reich.

Germany had spent billions in today's dollars to fund fundamental and applied scientific research to give the Nazi war machine a strategic advantage developing secret advanced weaponry, including jet airplanes and rockets capable of delivering bombs. The V-2 rockets hitting London made international headlines. Much less appreciated was that German scientists had cracked the chemical code unlocking the secrets of how petroleum products are formed. Starting in the early part of the twentieth century, German chemists developed the formulas necessary to produce synthetic oil. While the goal was to make gasoline, diesel fuel, and aviation fuel from Germany's abundant coal supply, the equations in what came to be known as the "Fisher-Tropsch" process explained the origin of oil as a naturally occurring phenomenon in which hydrogen and carbon bond, with ramifications far beyond turning coal into liquified synthetic fuel.

The Fisher-Tropsch Process

"Germany has virtually no petroleum deposits," observed Anthony N. Stranges of the Department of History at the Texas A&M University, noting a resource reality even today. "Prior to the twentieth century this was not a serious problem because Germany possessed abundant coal resources. Coal provided for commercial and home heating; it also fulfilled the needs of industry and the military, particularly the navy."[2]

But in the opening decade of the twentieth century, Germany's fuel requirements began to change. Germany became increasingly dependent on gasoline and diesel oil engines, as automobiles, trucks, and eventually airplanes made a plentiful supply of gasoline necessary. Then Germany's ocean-going ships, including the country's navy, converted from coal-burning to diesel oil as their energy source. "Petroleum was clearly the fuel of the future," Stranges noted, and Germany had a problem. Without ample petroleum resources, how was twentieth-century Germany going to develop the abundant gasoline and diesel fuel supplies

---

[2] Anthony N. Stranges, Department of History, Texas A&M Univerity, College Station, Texas 77843-4236, prepared for presentation at the AIChE 2003 Spring National Meeting, New Orleans, LA, March 30-April 3, 2001, unpublished.

needed to power a competitive national industrial economy and mount a second-to-none military operation in Europe?

The solution came in the 1920s in Berlin, when two German chemists, Franz Fischer (1877-1947) and Hans Tropsch (1889-1935), developed a series of equations that became known as the "Fischer-Tropsch Process," defining a methodology for producing synthetic gasoline and diesel fuel from coal. During the early 1930s, German industrial giant I.G. Farben received support from the Luftwaffe under Chancellor Adolph Hitler by proving the company could produce a high-quality aviation fuel. The army, the Wermacht, followed suit by lobbying to develop a domestic synthetic fuels industry. By 1936, I.G. Farben was no longer an independent company, but a government-private enterprise run by the Nazi high command.

Without the Fischer-Tropsch process, Hitler and Nazi Germany would have lacked the fuel resources needed to launch World War II. When Hitler attacked Poland on Sept. 1, 1939, Nazi Germany had fourteen synthetic fuel plants in full operation and six more under construction, producing approximately 95 percent of the aviation fuel used by the Luftwaffe. By 1943, using synthetic oil production defined by the Fisher-Tropsch process, Germany produced almost three million metric tons of gasoline by hydrogenation of coal. Adding to this diesel fuel, aviation fuel, and various lubricants produced synthetically from coal, Nazi Germany was able to satisfy up to 75 percent of its fuel demand through coal conversion processes made possible by the equations developed in the Fisher-Tropsch process.[3]

Imperial Japan, also constrained by lacking extensive national petroleum reserves, followed Nazi Germany into synthetic fuel production. In 1936, Japan calculated that the nation had a 400–500-year fuel reserve, if coal could be converted to liquid fuel. Japan's Seven Year Plan of 1937 called for the construction of eighty-seven synthetic fuel plants using the Fischer-Tropsch process by 1944, with the goal of producing 6.3 million barrels annually each of synthetic gasoline and synthetic diesel fuel. While the economic demands of waging war in China and across the Pacific ultimately thwarted Japan's ambitions to produce synthetic oil, Japan constructed fifteen synthetic fuel plants that reached peak production of 717,000 barrels in 1944.[4]

Operation Paperclip: US Military Intelligence Grabs Nazi Oil Secrets

While US Army intelligence officers had the first jump at confiscating Nazi scientific documents and interviewing Nazi scientists, by 1948, British intelligence, Canadian intelligence, and Russian intelligence all joined in, focusing their efforts to

---

[3] "The German Document Retrieval Project," Center for Energy & Mineral Resources," Texas A&M University, Sept. 20, 1977.

[4] Paul Schubert, Steve LeViness, and Kym Arcuri, Syntroleum Corporation, Tulsa, OK, and Anthony Stranges, Texas A&M University, "Fischer-Tropsch Process and Product Development During World War II," April 2, 2011, unpublished paper, at Fisher-Tropsch.org, under "Primary Documents/Presentations."

understand how the Nazis had produced synthetic petroleum products so successfully.

Ultimately, under the auspices of "Operation Paperclip," the Office of Strategic Services, or OSS (the predecessor agency to the CIA), hundreds of Nazi scientists and engineers were secretly brought to the United States. Many Nazi scientists were allowed to enter the United States despite their complicity in some of the Nazis' most horrific war crimes, including using political prisoners from the Holocaust as guinea pigs in terrifying "scientific experiments" and employing Jews and other political prisoners as slave labor in Nazi war-machine factories.[5]

An examination of the now declassified Operation Paperclip files at the National Archives and Records Administration in Washington, D.C., documents that a total of seven German synthetic fuels scientists were brought stateside, including the two most prominent then alive: Helmut Pichler and Leonard Alberts.

Helmut Pichler

Pichler, born on July 13, 1904, in Vienna, Austria, was forty-one years old when World War II ended in Europe. He worked as Franz Fischer's research assistant at the Kaiser Wilhelm

---

[5] See: Linda Hunt, Secret Agenda: The United States Government, Nazi Scientists, and Project Paperclip, 1945 to 1990 (New York: St. Martin's Press, 1991); and Tom Bower, The Paperclip Conspiracy: The Hunt for the Nazi Scientists (Boston: Little, Brown and Company, 1987).

Institute, perhaps Germany's most prestigious pre-war scientific institution. When interviewed by the Office of the US Military Government at the end of the war in Germany, Pichler had published fifty scientific articles and held nineteen patents on a wide range of topics related to the chemistry and manufacturing of synthetic fuels. In his biographical and professional data debriefing with US military intelligence, Pichler boasted he was "co-inventor" of the benzene-synthesis process from which synthetic gasoline was produced. At the end of the war, Fischer was approaching seventy years old. But Pichler, undoubtedly one of the most knowledgeable and accomplished synthetic fuels scientists in the world, was still young enough to travel and continue advancing his professional career.

Pichler's file contains a letter from Fischer himself, dated June 23, 1947, writing as director of the Kaiser Wilhelm Institute of Coal Research, a position Fischer held from 1913 until 1943. Pichler joined the institute in March 1927, and two year later completed his doctoral thesis on the subject, "About the Synthesis of Hydrocarbons." After his graduation, Pichler was Fischer's assistant until April 1936, when Pichler was appointed the head of the division for synthetic fuels. Pichler was then nominated to become a permanent scientific member of the Kaiser Wilhelm Institute of Coal Research.

In the letter, Fischer credits Pichler with a long list of scientific accomplishments, including developments in the field of the synthesis of gasoline, research using both iron and cobalt as catalysts in the Fischer-Tropsch synthetic fuel production process, and the conversation of methane to more complex hydrocarbon chains, including benzene and acetylene. "The work of Dr. Pichler has contributed substantially to the technical scale development of the normal-pressure-synthesis of Franz Fischer and Hans Tropsch after Dr. Tropsch left the Institute [in 1926]," Fischer's letter continued. "Fundamentally separate, Dr. Pichler developed the mentioned medium-pressure-synthesis, the high-pressure synthesis of paraffins and the other topics mentioned above." Fischer, writing in English, concluded his letter with an unqualified endorsement: "Dr. Pichler was one of the best co-workers I ever had. His personal qualities are the factors for which not only the scientific, but also the social intercourse with him were very pleasing in the 16 years of our cooperation."

Pichler's signed "Statement Concerning Past Political Affiliations," submitted as part of his interrogation by US military intelligence, indicates that in 1932, Fischer urged him to become a citizen of Germany. In 1933, Pichler became a member of the Nazi Party. In 1934, at the request of the SA, he gave ten lectures concerning air defense, including how to fight incendiary bombs, although he professed to do so out of fear of reprisals, not for any enthusiasm to be involved for political reasons. "All my thoughts and my sympathies were ever concerned with my scientific work only," he wrote in his signed statement. "I performed this work in the same way before 1933, after 1933 and after 1945." He claimed he wanted to come to the United States to continue his scientific research and to become a US citizen.

The Truman administration was sufficiently enthusiastic to get a synthetic fuel scientist with Pichler's credentials to come to the United States that Pinchler was given the benefit of the doubt that his Nazi affiliations were more a matter of necessity that political preference. Also overlooked was Pichler's major contribution to the Nazi war effort, given how important the production of synthetic fuels was to the Nazis' ability to wage war. The US government gave Pichler permission to enter the country along with his wife Louise Maria, then forty-four years old, and his two daughters, Christa, age eleven, and Irmstraud, age five, as well as his son, Rolf-Helmut, age ten.

Once in the United States, Pichler joined Hydrocarbon Research Inc., where he helped construct a commercial Fischer-Tropsch plant in Brownsville, Texas. In his later years, Pilcher was quoted as saying that German scientists and engineers interviewed by US intelligence operatives at the end of World War II did not divulge all they knew. The truth is that up until 1940, German scientists and engineers, with the consent of the Nazi government, had been providing technical information about the Fischer-Tropsch process to a consortium of six companies that had been members of the old Standard Oil Company. Beginning in 1938 and 1939, Standard Oil also began purchasing common stock of

Hydrocarbon Research, Inc.[6] Records show that Standard Oil and I.G. Farben in Germany had been interested in and cooperating regarding synthetic fuels since the twenties and thirties.

Leonhardt Alberts

In contrast to Pichler, Leonhardt Alberts was so enthusiastically a Nazi even after World War II ended that it required a US government cover-up to get him clearance to enter the United States over political objections voiced by top officials in the Truman administration.

Alberts was five years older than Pichler. He was forty-six years old at the end of the war with Germany, born on April 21, 1899, in Oanabrueck, Germany. He was the plant manager and technical director of Ruhchemie, A.G. (the Ruhr Chemical Corporation in Oberhausen, Rhineland) from 1929-1943. Then, from 1943 through 1946, he was on the board of directors of synthetic nitrogen and hydrocarbon plants for Victor Works, in Castrop-Rauxel, Germany. At the end of the war, there was no one in Germany more expert at operating and managing synthetic fuel plants than Leonhardt Alberts.

The problem was that Alberts was a candidate for the Nazi party as early as 1933, joining formally in 1938. Subsequently, he

---

[6]   Burton H. Davis, Center for Applied Energy Research, University of Kentucky, 2540 Research Park Drive, Lexington, KY, 4-511, "An Overview of Fischer-Tropsch Synthesis at the US Bureau of Mines," prepared for presentation at the AIChE 2003 Spring National Meeting, New Orleans, LA, March 30-April 3, 2001, unpublished.

belonged to both the SS and the SA. The Operation Paperclip file even preserved two yellow-page legal pads with the handwritten notes taken by the FBI agent conducting a background search on Alberts. The research leaves no doubt that Alberts was an ardent Nazi, even after he had received permission from the US government to immigrate to the United States along with his family.

"Mr. H.T. McBride, Projects Supervisor, Bechtel Corporation where Alberts was ultimately hired, related that his associations with Alberts have been entirely disagreeable," the FBI case file for Alberts noted. "During his stay here, Alberts exhibited an arrogant and domineering attitude in regard to company administrative matters. He was non-cooperative in obeying regulations pertaining to expenses of travel, leave arrangements, and the certification of time off, to name a few. In the opinion of Mr. McBride, Alberts is exceedingly ambitious, and will try every trick and scheme he knows which might work to his sole benefit." McBride told the FBI he believed Alberts was "a true Nazi" and "wholly undesirable for citizenship," and that he felt admitting Alberts to the United States "would be a definite threat to the security of this country."

C. W. Frye, personal manager at the Bechtel Corporation, gave the FBI a similar report. Frye said he had "no sympathy" for Alberts' desire to become citizen of the United States. He characterized Alberts as "non-cooperative and disagreeable almost without exception in business contacts." He charged that Alberts "has an overbearing demeanor which appears to be self-trained." Frye advised the FBI that Alberts "has few of the qualities necessary to becoming a good citizen," as he concluded Alberts was not a good security risk.

Major Robert E. Humphries, Quartermaster Corps, US Army, agreed. Humphries told the FBI that Alberts is "poorly regarded" because of "his insufferable and pompous attitude." Humphries commented that Alberts "certainly never exhibited any remorse or sense of guilt arising out of his past connections in Germany," and he charged Alberts "was and is a Nazi." Humphries further advised that Alberts "would be a dangerous man" to admit into the United States as a permanent resident because Alberts would be given an ample opportunity to learn all details of the synthetic fuels program in this country. At the same time, he was distrustful of allowing Alberts to return to Germany as a free man because he believed he would be capable of "dealing with Russia or with any other group which would pay for his technical knowledge."

Alberts argued in a signed "Political Bibliography" written at the request of US military intelligence and included in his Operation Paperclip file what appears to be a self-serving explanation that he had joined the Nazi party for political expediency only:

> As Director of the Ruhrchemie A.G. in 1933, I
> was naturally pressed to affiliate myself with the
> N.S.D.A.P. [the Nazi Party]. It was possible for me

in contrast to the other Directors of my firm to keep aloof from this membership.

In 1935 I was offered the position on the Board of Directors of the Briunkohle-Benzin A.G. However, after it had been determined that I was not a member of the N.S.D.A.P., this offer was withdrawn. In 1938 I got a similar offer from Krupp. This offer was also withdrawn for the same reason.

After two examples convinced me that without party membership I would not be able to accept offers which would improve my professional position. Therefore, I applied for a membership in 1938.

On Nov. 9, 1949, Peyton Ford, the assistant to the US attorney general, wrote to Colonel Daniel E. Ellis, US Air Force, and director of the Joint Intelligence Objectives Ageny in the Pentagon, to urge that Alberts' continued presence in the United States represented a risk to internal security:

Upon consideration of all the information received concerning Alberts this Department is of the opinion that it cannot recommend him to the Immigration and Naturalization Service for permanent admission into the United States. You still note that Alberts served for a time during World War II as a functionary of the Abwehr, the German Intelligence. The statements of several persons who have known Alberts, including Major Robert E. Humphries, who has been directly concerned with security matters pertaining to the presence of German scientists at Bureau of Mine plants, have grave misgivings of Alberts as a security risk. It would appear that he is a pro-Nazi in his outlook and unscrupulous in his activities and, as Major Humphries has stated, he is capable of dealing with Russia or any other group which would pay for his technical knowledge.

What ensued was a bureaucratic fight within the government between those that coveted Alberts' technical skills, and those charged with policing security risks. Acting Secretary of Commerce Thomas C. Blaisdell weighed in strongly favoring Alberts, dismissing the security concerns as unimportant.

In a letter to the Attorney General J. Howard McGrath, dated July 14, 1950, Blaisdell stated, "The Fischer-Tropsch process for the production of synthetic fuels, in which Albert is expert, may be a significant item in our national defense."

On Feb. 24, 1949, H.H. Storch, research and development branch chief of the Office of Synthetic Liquid Fuels in the US Department of Interior's Bureau of Mines, wrote to the Department of Commerce and offered more support, praising the work Alberts had done consulting on Fischer-Tropsch pilot plant work. Storch had written:

During Mr. Alberts' stay under the direction of the Bureau of Mines, he contributed to the development of a process which originated in Germany at the I.G. Farbenindustrie, and which was being completed by the Bureau of Mines. We found him to be a good, practical engineer. His character and general behavior were excellent and, so far as we can tell from our observation of him at work, he would make a good citizen of the United States.

The Operation Paperclip files show the commercial interests within the government won out and Alberts was given permission to enter the United States along with his wife, Agnes, his sister and his sister-in-law.

Post-War Synthetic Fuel Plants in the United States

In 1949, the Bureau of Mines opened a synthetic fuels demonstration plant in Louisiana, Missouri, on 390 acres of a former war department ammonia plant that was located seventy-five miles north of St. Louis. Bechtel operated this $10 million coal hydration plant, with some 400 employees that included the seven Nazi synthetic fuel scientists Operation Paperclip brought to the United States.

From 1950 to 1952, Hydrocarbon Research Inc. built and operated a synthetics fuel plant in Brownsville, Texas. The

Bureau of Mines conducted numerous synthetic fuel pilot projects, none of which reached commercial viability.

While the post-war efforts of the US government to develop synthetic fuel plants were successful, the project never took root in a global economy where the production of petroleum "fossil fuels" was both abundant and commercially profitable. Put simply, US oil companies had no reason to develop relatively expensive synthetic oil when billions of dollars in profits could be made annually bringing to market naturally produced and reasonably priced hydrocarbon fuels, including crude oil and natural gas. The production of synthetic fuels, while technically interesting to US oil companies and government officials, was considered too costly to pursue commercially when traditional crude oil and natural gas reserves available on the global market were still relatively abundant and reasonably cheap to discover, develop, and bring to market in the United States.

By the 1960s, the US government interest in synthetic fuels was largely academic. Taxpayer funding for Fischer-Tropsch research dried up, and work the US Bureau of Mines undertook in the postwar period was transferred in the 1960s to the Office of Coal Research in the Department of the Interior, and then in the 1970s to the Energy Research and Development Administration. In 1977, Congress created the US Energy Department, and the public policy emphasis shifted away from synthetic fuels to the "fossil fuel" program. On June 30, 1980, the Energy

Security Act was signed into law, creating the United States Synthetic Fuels Corporation that would provide financial assistance to the private sector to stimulate production of synthetic fuels. Only one plant was ever actually built.[7]

As a result of the public policy emphasis on utilizing abundant "fossil fuel" resources, the Nazis' petroleum secrets languished. Hundreds of thousands of pages from confiscated German scientific papers on the Fisher-Tropsch remained classified until the late 1970s. In Oct. 1975, the Texas A&M University's Center for Energy and Mineral Resources initiated a project to locate, retrieve, abstract, and index the German World War II industrial records with the objective to publicize the processes Nazi Germany had used to produce synthetic fuel. By 1977, the twelve full and part-time members of the project staff brought to Texas A&M 310,000 pages of documents, consisting primarily of the 305 Technical Oil Mission microfilm reels and twenty-five microfilm reels collected by Air Force Intelligence at the end of World War II.

But, even today, countless thousands of pages of Fischer-Tropsch scientific studies confiscated from Germany at the end of World War II lie deteriorating, never translated, in aging and neglected paper and microfilm archives. Remarkably, despite the efforts of Texas A&M and the National Archives, the process of locating confiscated Nazi synthetic petroleum documents for scientific study remains difficult, if not virtually impossible even

---

[7] Ibid.

17

today. When found, most of the documents remain as they were when first confiscated in 1945—never so much as summarized or abstracted in English, let alone translated in full. On Sept. 20, 1977, the German Document Retrival Project concluded the following: "Knowledge in these [German] documents [on synthetic fuels] has for all practical purposes not been available to industry, government, educational institutions or the public at large."[8]

Over time, the Fischer-Tropsch process was relegated to the point where those aware of the Fischer-Tropsch process though of it primarily as a way of liquefying coal to produce gasoline and diesel fuel. Why bother liquefying coal when the US still had abundant oil and natural gas reserves available domestically or on international markets at a relatively reasonable price? Even in oil crises, such as the 1975 OPEC oil embargo under President Jimmy Carter, few serious politicians or scientists thought seriously about reviving interest in synthetic liquid fuel.

Today, few Americans know anything about the World War II achievements of the Nazis in developing synthetic fuel. How different this is from the enthusiasm of the US military's Technical Oil Mission, which at the end of World War II had targeted all Nazi synthetic fuel plants, including refineries and chemical plants, all research laboratories, including the Kaiser

---

[8] "The German Document Retrieval Project," loc.cit, supra at note #3.

18

Wilhelm Institute, and corporate headquarters, such as I.G. Farben.

Decades after the end of World War II, US petro-scientists and petro-geologists remain locked in the vision that the only productive petroleum science and geology derives from an understanding that oil and natural gas are biologically produced "fossil fuels." Rather than study the Fisher-Tropsch equations to unravel the code of how hydrocarbons are produced, US petro-scientists and petro-geologists are happy to relegate those Nazi documents to obscurity because they consider synthetic oil production a waste of time.

Nazi synthetic oil secrets remain hidden from the public view because that's exactly the way US oil companies and the US government want it. The true secret of Nazi synthetic oil has nothing to do with liquefying coal. Perhaps this was what Helmut Pichler had in mind when he said that Nazi scientists never completely revealed every secret their explorations of synthetic fuels unveiled. Committed Nazis such as Leonhard Albert might have been quietly pleased when American scientists saw nothing more in the Fischer-Tropsch process than how to make gasoline and diesel fuel out of coal.

What the German synthetic fuel scientists truly cracked was the code God built into the heart of chemistry to form hydrocarbons in the first place. Beyond the formulas to make gasoline and diesel fuel out of coal, what the Fischer-Tropsch process postulates is that hydrocarbons form naturally in the mantle of the earth on an on-going basis. The Fischer-Tropsch equations reveal the formulas through which compounds including hydrogen and compounds including carbon, in the presence of a catalyst such as iron ore or cobalt, could form hydrocarbon chains under conditions of extreme heat and pressure known to exist in the mantle of the earth. Applying this knowledge to making gasoline and diesel fuel from coal served the Nazi war machine in a country lacking readily available hydrocarbon resources close to the surface of the earth. Revisiting the Fischer-Tropsch equations today is important not specifically to develop synthetic fuel, but to present a direct challenge to the fossil fuel theory of the origin of oil. The Fischer-Tropsch equations understood in the context of fundamental scientific research, not just applied scientific research, challenge the concept that oil and natural gas have an organic origin. Ironically, what remains even today locked away within the Fischer-Tropsch equations is an understanding that all hydrocarbon fuels are abiotic in origin, produced naturally in the mantle of the earth on a continuous basis, without the involvement of any organic material whatsoever.

Russia and Deep-Earth Oil

The truth is that only Soviet Russia under dictator Joseph Stalin truly benefited from the confiscated Nazi petroleum secrets.

On Nov. 3, 1944, well before the end of the war, President Franklin Roosevelt issued a directive calling for a government study to answer the following question: What precisely did dropping over 2.7 million tons of bombs on Europe accomplish?[9]

The resulting *United States Strategic Bombing Study* produced some surprising results. The attack on the German airplane industry culminated in the last week of Feb. 1944, when 3,636 tons of bombs were dropped on airframe plants. Every known aircraft factory in Germany was hit. But, surprisingly, in 1944 the Nazis accepted a total of 39,807 aircraft of all kinds, when the number accepted in 1942 before the bombing attacks began had only been 15,596. The German aircraft production had actually increased despite the massive bombing of Nazi aircraft plants.

Why? The bombing destroyed the buildings, but the machines "showed remarkable durability." The Germans reorganized the management of the aircraft plants and subdivided production into many small units that were immune to massive bombing raids. As the aircraft manufacturing plants were being destroyed, the Germans adapted, learning how to recover the machinery and disperse the manufacturing. The result was clear – bombing the plants had not slowed down the Nazis' ability to make new airplanes.

The allied bombing of German oil and chemical production plants told a different story. By the end of the war, the

---

[9] The United States Strategic Bombing Survey. The European War report was the first completed, published by the Government Printing Office on September 30, 1945. This report as originally issued can be read on the Internet at the following URL: http://www.anesi.com/ussbs02.htm#page1.

Germans could produce fighter planes and bombers, but they had no airplane fuel with which to fly the aircraft. The output of aviation gasoline from synthetic plants fell in Nazi Germany from 316,000 tons per month, when the air attacks began in 1943, to 5,000 tons in Sept. 1944, when every major synthetic fuel plant in Germany had been bombed from the air by the Allies. Without fuel, the Nazi war machine came to a grinding halt.

Once the war was over, Stalin determined that the Soviet Union would never be vulnerable to a foreign enemy because of a dependence on foreign oil. He resolved that Russia would become oil self-sufficient, as part of his plans for expanding communism and Soviet domination worldwide. US petro-scientists and petro-geologists looking for oil as "fossil fuel" formed in sedimentary rock structures lying relatively close to the surface of the earth concluded that Russia, like Germany, lacked petroleum reserves. But Stalin ordered his petro-scientists to study the Fisher-Tropsch process, anxious to see if what the Germans understood about the origin of oil might help Russia become energy independent, regardless of what the American scientists said.

Beginning in 1940, Stalin commissioned a scientific examination into every aspect of petroleum, including how it is created, why reserves are formed, and how the oil can best be discovered and extracted from the earth. Between 1940 and 1995, Russian scientists published 347 scientific publications on the Fisher-Tropsch process, and obtained some 170 Fisher-Tropsch patents.[10] By 1951, Professor Nikolai Kudryavtsev articulated what today has become known as the Russian-Ukranian Theory of Deep, Abiotic Petroleum Origins. The theory rejects the contention that petroleum products are formed from the remains of ancient plant and animal life that died millions of years ago, arguing instead that petroleum products are abiotic in origin.

According to Professor Kudryavtsev, oil has nothing to do with living organisms rotting into petroleum. The Soviet scientist ridiculed the idea that an ancient primeval morass of plant and animal remains was covered by subsequent millions of years of sedimentary deposits, only to be compressed by the millions of more years of heat and pressure. The Soviet theory as advanced by Kudryavtsev and dozens of Russian scientists who followed him was that the origin of oil was "a-biotic." In other words, oil did not come from the once-alive, "biotic" material of ancient plants and animals. Instead, the Soviet scientists concluded that the chemical processes by which hydrocarbons were produced were a natural product of the earth itself, manufactured at deep levels where there never were any plants or animals. Abundant oil could be found in Russia, the Soviets concluded, if only wells were drilled deep enough.

---

[10] V.I. Anikeev, Y. Yermakova, B.L. Moroz, Boreskov Institute of Catalysis, Novosibirsk, Russia, "The State of Studies of the Fischer-Tropsch Process in Russia," unpublished paper supported by Syntroleum Corporation, Tulsa, Oklahoma.

Today, contrary to the predictions of US petro-scientists and petro-geologists at the end of World War II, Russia rivals Saudi Arabia as the world's leading producer of crude oil.

Just to be clear, please understand that the argument here is that all oil and natural gas produced by the earth are abiotic in origin. Please do not think that the oil and natural gas US petro-scientists and petro-geologists, thinking traditionally, find in sedimentary rock are organic in origin, while only oil and natural gas found at deep-water or deep-earth levels are abiotic. Granted, synthetic fuels can be formed from a wide variety of organic substances, ranging from corn and sugar cane, to animal parts, and even sewage. Generally, the synthetic processes used to transform organic material into synthetic fuel involve well-understood chemical transformations very similar to the fermentation and bacteriological processes commonly used to transform organic materials into various alcoholic beverages. The argument here is that oil and natural gas found near the surface of the earth in sedimentary rock structures were formed at deep-earth levels and poured through cracks in the earth's bedrock structure, only to pool in the more porous sedimentary rock structures typically found near the earth's surface.

The point is that fuels produced from organic material generally require human action to be formed. Hydrocarbon fuels produced naturally by the earth are never "fossil fuels" produced through biologic materials; they are always abiotic in nature. Just as fossils are never the ancient flora or fauna themselves, there are no "fossil fuels," regardless what petro-scientists and petro-geologists teach college students in university classrooms.

In nature, hydrogen and carbon do not require the intervention of any dead or living material. Instead, all nature requires to produce petroleum are the synthetic fuel equations the German chemists developed, beginning with Franz Fischer in the 1920s and culminating with the Fischer-Tropsch synthetic fuel plants operated by the Nazis during World War II.

2

The Suppressed
Science of Abiotic Oil

In the United States, the abiotic theory of the origin of oil is still ridiculed as "a conspiracy theory" by a scientific community wedded to the concept that oil is produced by organic material. Most geo-scientists have at least advanced to the point where the idea that buried dinosaurs and ancient forests produce oil is considered ridiculous. Yet the idea that oil derives from ancient biological debris persists. Hydrocarbon energy is still considered "fossil fuel," even though by definition a fossil is not the actual animal or plant itself, but the structure of the animal or plant typically filled in by various minerals that harden into stone over the ages. Despite this, the vast majority of US geo-scientists find it impossible to imagine that oil can have anything but a biological origin, such that the politically correct scientific consensus remains even today that organic materials such as plankton and algae are responsible for creating oil.

How Exactly Do "Fossils" Make "Fuel"?

What then is the supposed chemical processes by which decaying plants and dinosaurs, or plankton and algae, are supposed to decay into "fossil fuel"?

Richard Heinberg, a senior fellow-in-residence at the Post Carbon Institute in Santa Rosa, California, has argued that "the assertion that all oil is abiotic requires extraordinary support, because it must overcome abundant evidence" that ties "specific oil accumulations to specific biological origins through a chain of well-understood processes that have been demonstrated, in principle, under laboratory conditions."[11] So, if what Heinberg asserts is true, we should have no problem discovering the precise laboratory-proven formula under which biological material decays into hydrocarbon fuel.

Seppo Korpela of the Ohio State University Department of Mechanical Engineering argues that fossil fuels form when "the early sedimentary layers" at the bottom of a basin are deprived of oxygen such that the organic matter in them does not decay, "as it does in the common setting of kitchen compost."[12] Then, "anaerobic bacteria" can "go to work and turn the organic material into the substance kerogen." Kerogen, Korpela posits, can be thought of as "immature oil." The term "anaerobic" refers to a process occurring in the absence of

---

[11] Richard Heinberg, "The 'Abiotic Oil' Controversy," Energy Bulletin, Oct. 6, 2004, at http://energybulletin.net/node/2423.

[12] Seppo A. Korpela, "Oil Depletion in the United States and the World," a working paper for a talk to Ohio Petroleum Marketers Association at their annual meeting in Columbus, Ohio, May 1, 2002, at http://greatchange.org/ov-korpela,US_and_world_depletion.html.

free oxygen. When kerogen is found at depths of between 6,000 and 13,000 feet and when the temperature and pressure are "right," the kerogen "in the *source rock* will be cracked into oil. This zone is called the *oil window*. At depths greater than 13,000 ft. temperatures are so high that oil is cracked into gas."

"Kerogen," it turns out, is not a chemist's term. Kerogen is a loose, geological term (deriving from the ancient Greek word *keros*, meaning wax) that an industry oil glossary defines as: "The naturally occurring, solid, insoluble organic material that occurs in source rocks and can yield oil upon heating."[13] Kerogen is not a term typically found in chemistry textbooks or specifically used by professional chemists. Use of the term "kerogen" is generally a signal the person is a petroleum geologist, not a chemical scientist.

Ker Than, a staff writer for LiveScience.com, provides the common sense explanation for how kerogen is supposed to transform into "fossil fuel":

> *In the leading theory, dead organic material accumulates on the bottom of oceans, riverbeds or swamps, mixing with mud and sand. Over time, more sediment piles on top and the resulting heat and pressure transforms the organic layer into a dark and waxy substance known as kerogen.*

---

[13] "Kerogen," in the Schlumberger Oilfield Glossary, at http://www.glossary.oilfield. slb.com/Display.cfm?Term=kerogen.

Left alone, the kerogen molecules eventually crack, breaking into shorter and lighter molecules composed almost solely of carbon and hydrogen atoms. Depending on how liquid or gaseous this mixture is, it will turn into either petroleum or natural gas.[14]

Chemical textbooks typically do not provide chemical formulae for kerogen.

The transformation from "kerogen" to "fossil fuels" appears to be more a matter of faith than an observed process that can be described in a precise chemical formula and replicated in a laboratory.

Published scientific analyses attempting to describe "the notion of kinetic cracking of kerogen into petroleum" tend to start by pointing out that the explanation is not particularly rigorous. According to M. Vandenbroucke of the French Institute of Petroleum, "It is important to keep in mind that the name kerogen, in opposition with usual chemical nomenclature, does not represent a substance with a given chemical composition. Indeed kerogen is a generic name, in the same sense as lipids or proteins."[15]

Technical discussions of how kerogen produces oil from source rock generally end up describing field-oven heating devices typically designed to analyze rock samples, such as

---

[14]   Ker Than, "The Mysterious Origin and Supply of Oil," *Live Science,* Oct. 10, 2005, at http://www.livescience.com/9404-mysterious-origin-supply-oil.html.

[15]   M. Vandenbroucke, "Kerogen: from Types to Models of Chemical Structure," *Oil & Gas Science and Technology,* Rev. IFP, Vol. 58 (2003), No. 2, pp. 243-269, at http://ogst.ifpenergiesnouvelles.fr/index.php?option=com_article&access=standard&Itemid=129&url=/articles/ogst/pdf/2003/02/vandenbroucke_v58n2.pdf.

the Rock-Eval prolysis device, into which geologists can cook "source rock" in the field to see if the specimen rock looks like other "source rock" where oil has already been found.[16] Again, the result is practical field geology, not rigorous laboratory science specifying chemical formulae identifying how flora and protoplasm turn into hydrocarbons.

Still lacking are the laboratory demonstrations authors such as Richard Heinberg claimed we would find. Scientific literature written by petro-geologists and petro-chemists is typically lacking in rigorous scientific experiments specifying the equations to demonstrate precisely how rock turns into kerogen and subsequently into hydrocarbon fuel, suggesting the entire concept may be better classified as alchemy.

Methane Synthesized in a Laboratory

In 2004, Henry Scott of Indiana University in South Bend, organized a research team including Dudley Herschbach, a Harvard University research professor and recipient of the 1986 Nobel Prize in chemistry, as well as scientific colleagues from Harvard University, the Carnegie Institute in Washington, and the Livermore National Lab, to see if they could synthetically

---

[16] See, for instance: M. Teichmüller and B. Durand, "Fluorescence microscopical rank studies on liptinites and vitrinites in peat and coals, and comparison with results of the rock-eval pyrolysis," *International Journal of Coal Technology*, Vol. 2, Issue 3, February 1983, pp. 197-230, at http://www.sciencedirect.com/science/article/pii/0166516283900010.

produce methane in a laboratory without using organic materials of any kind.[17]

The research team decided to squeeze together iron oxide, calcium carbonate, and water at temperatures as hot as 500 degrees Celsius and under pressures as high as 11 gigapascals (one gigapascal is equivalent to the pressure of 10,000 atmospheres). Simply put, the scientists were testing a fundamental principal of the Fischer-Tropsch equations, trying to see if the combination would produce methane if exposed to pressures and temperatures comparable to those experienced in the earth's upper mantle.

To conduct the experiment, the scientists designed a "diamond anvil cell" mechanism consisting of two diamonds, each about three millimeters high (about one-eighth inch). The tips of the diamonds were pointed together, allowing them to compress a small metal plate designed to hold the sample of iron oxide, calcite (the primary component of marble), and water that the scientists wanted to force together. The scientists then conducted a variety of highly accurate spectroscopic analyses on the sample material that resulted. Herschbach explained that diamonds were ideal material for the experiment because, as

---

[17] Henry P. Scott, Russell J. Hemley, Ho-kwang Mao, Dudley R. Herschbach, Laurence E. Fried, W. Michael Howard, and Sorin Bastea, "Generation of methane in the Earth's mantle: In situ high pressure-temperature measurements of carbonate reduction," Proceedings of the National Academy of Sciences of the United States of America, Vol. 101(39), September 28, 2004, pp. 14023-14026, at http://www. ncbi.nlm.nih.gov/pmc/articles/PMC521091/. Also, see: http://www.pnas.org/ content/101/39/14023.full.pdf+html.

one of the "hardest substances on earth, they can withstand the tremendous force, and because they're transparent, scientists can use beams of light and X-rays to identify what's inside the cell without pulling the diamonds apart."[18]

The basic idea was to smash the iron oxide, calcite, and water together at the types of temperatures and pressures we would expect to see deep within the earth and stand back to see what happened. The diamond mechanism provided to be a reliable way to take the end product and submit it to spectrographic analysis so its chemical content could be analyzed accurately. The goal was to prove that a hydrocarbon of the petroleum family could be produced via simple inorganic reactions involving no biological agents whatsoever.

Remarkably, the experiment worked. The scientists found they could easily produce methane, the principal component of natural gas, at temperatures around 500 degrees Celsius and at pressures of seven gigapascals or greater. Inorganic chemicals (iron oxide, calcium carbonate, and water) had been combined to produce the "organic" chemical, methane. Laurence Fried of Livermore Laboratory's Chemistry and Minerals Science Directorate summed up the importance of these findings as follows:

> The results demonstrate that methane readily forms
> by the reaction of marble with iron-rich minerals and
> water under conditions typical in Earth's upper man-

---

[18]   Quoted in: Erin O'Donnell, "Rocks into Gas," *Harvard Magazine*, March-April 2005, at http://harvardmagazine.com/2005/03/rocks-into-gas.html.

tle. *This suggests there may be untapped methane reserves well below Earth's surface.*

Dr. Fried continued:

> *At temperatures above 2,200 degrees Fahrenheit, we found that the carbon in calcite formed carbon dioxide rather than methane. This implies that methane in the interior of Earth might exist at depths between 100 and 200 kilometers. This has broad implications for the hydrocarbon reserves of our planet and could indicate that methane is more prevalent in the mantle than previously thought. Due to the vast size of Earth's mantle, hydrocarbon reserves in the mantle could be much larger than reserves currently found in Earth's crust.*[19]

The research further showed that methane is thermodynamically stable under conditions typical in the mantle of the earth, "indicating that such reserves could potentially exist for millions of years."[20] Moreover, the scientists concluded "the potential may exist for the high-pressure formation of heavier hydrocarbons by using mantle-generated methane as a

---

[19] Quoted in: DOE/Lawrence Livermore National Laboratory, "Methane in deep earth: A possible new source of energy," Press Release, Energy Bulletin, Sept. 12, 2004, at http://www.energybulletin.net/node/2093.

[20] "An Inexhaustible Source of Energy from Methane in Deep Earth," Psysorg.com, Sept. 15, 2004, at http://www.physorg.com/news1166.html.

precursor."[21] If methane could be generated synthetically in a lab, the scientists seemed to suggest, it could be a precursor to forming heavier hydrocarbons, possibly even petroleum, from abiotic processes in the earth's mantle.

In 1828, German chemist Friedrich Wöhler synthetically created urea by heating cyanic acid and ammonia. In other words, urea, then known only as an organic substance isolated from metabolically generated urine, had been generated by the combination of inorganic chemicals. This broke the presumption that "organic" chemistry was devoted to a "living" class of chemicals that resulted from and possibly contained a "vital life force." In a similar fashion, if methane can be created synthetically from inorganic chemicals, biological content is not a necessary a requirement to form hydrocarbons. Laboratory-produced abiotic methane directly challenges the theory that hydrocarbon fuels are by definition organic in origin. While this experiment generated only methane, and not the more complex hydrocarbon structures required for petroleum, the scientists involved stated their conclusion that their results encouraged them to believe that the more complex hydrocarbon structures could also be created in an abiotic manner.

In discussing the experiment, Herschbach noted that he derived inspiration from two previous thinkers: Dmitri Mendeleev, the nineteenth-century Russian scientist who invented the periodic table of elements, and Thomas Gold, the Austrian-born Cornell University astrophysicist perhaps

---

[21] Henry P. Scott, op.cit.

most responsible for introducing the idea of abiotic oil to a United States audience. Mendeleev, in 1877, was one of the first scientists to argue that petroleum is "born within the depths of the Earth, and it is only there that we must seek its origin."[22]

Thomas Gold: The Deep, Hot Biosphere

Thomas Gold was a professor of astronomy who taught at Cornell University and died in 2004, at eighty-four years old. In 1998, when he was 78, he published a controversial book entitled *The Deep Hot Biosphere: The Myth of Fossil Fuels*.[23] With this book, Gold ventured into geology, taking up the controversial position that suggested the Russian-Ukranian deep, abiotic theory on the origin of oil was right, despite being ignored by Western scientists and geologists.

Gold was born in Vienna in 1920 and studied in Switzerland before leaving for Cambridge University shortly before World War II broke out. For a year, Gold was held in a British internment camp, suspected of being an enemy spy. After being released , Gold helped develop radar for the British Admiralty before ending up in the United States, first at Harvard before he was recruited in 1959 by Cornell University, where he eventually chaired the astronomy department and directed the Center for Radiophysics and Space Research. He

---

[22] Dimitri Mendelev, "L'Origine du pétrole," Revue Scientifique, Second Series, VIII, p. 409-416.

[23] Thomas Gold. *The Deep Hot Biosphere. The Myth of Fossil Fuels* (New York: Copernicus Books, 1998). First softcover edition published by Copernicus Books in 2001.

had to wait until 1969 to get his doctorate, when Cambridge University finally decided to bestow upon him an honorary degree.

As an astronomer, Gold was well aware that hydrocarbons are abundant in our solar system. Since the early part of the twentieth century, spectrographs that analyze wavelengths have shown with certainty that carbon is the fourth most abundant element in the universe, right after hydrogen, helium, and oxygen. Furthermore, among planetary bodies, "carbon is found mostly in compounds with hydrogen – hydrocarbons – which, at different temperatures and pressures, may be gaseous, liquid, or solid. Astronomical techniques have thus produced clear and indisputable evidence that hydrocarbons are major constituents of bodies great and small within our solar system (and beyond)."[24]

In other words, hydrocarbons are not "organic chemicals" resulting from life processes on earth, as is commonly assumed by proponents of the fossil fuel theory. Rather, hydrogen is a fundamental element readily available in the universe, one that combines with carbon to form hydrocarbons, whether life is present or not. What astronomers have known about the abundance of abiotic hydrocarbons in the universe unfortunately has not passed over to geologists in the United States who typically think hydrocarbons form in the earth by organic processes.

What made sense to Gold was that hydrocarbons in various forms, including crude oil and methane gas, were some of the

---

[24]   Ibid., p. 44.

fundamental building blocks of earth as it formed and as it has continued to develop. Gold agreed with the Russian and Ukrainian scientists that petroleum is "abiogenic and ubiquitous deep in the earth."[25] In other words, go deep enough into the mantle of the earth, and you will find abundant oil everywhere. The reason we find oil in sedimentary rock is not because it is the "source rock" enclosing the rotting bio-matter, but because sedimentary rock is porous enough for the oil to pool into, and because fissures in the crust of the earth have permitted oil formed in the mantle of the earth to seep up and pool in sedimentary rock closer to the earth's surface.

Gold also postulated that at the bottom of the earth's oceans, hydrocarbons would seep from deep-water vents, providing gases and fluids needed for microbes to live, with no need of light or photosynthesis to provide nourishment for the microbes to flourish. He also argued that any presence of macrobiotic and bacterial life observed in petroleum reserves could have been picked from the layers of rock through which the oil passed on the way to the earth's surface. He concluded life is not confined to the surface of the planet. Instead, he saw earth itself as a biosphere, teeming with organisms living so deeply below the surface that the living organisms attaching to deep-water and deep-earth oil live, despite never having seen the light of day.

Interestingly, Thomas Gold also took a jab at scientists persisting in their convictions that oil has a biological origin, writing on page

---

[25] Ibid., page 39.

85 of *The Deep Hot Biosphere*: "Nobody has yet synthesized crude oil or coal in the lab from a beaker of algae or ferns."

Gold Confirmed: Abiotic Oil Found on Titan

NASA scientists, in conjunction with the European Space Agency and the Italian Space Agency, have determined from a Cassini-Huygens probe that first landed on Titan on Jan. 14, 2005, that the giant moon of Saturn contains abundant methane.

"We have determined that Titan's methane is not of biological origin, so it must be replenished by geological processes on Titan, perhaps venting from a supply in the interior that could have been trapped there as the moon formed," Dr. Hasso Niemann of the Goddard Space Flight Center told reporters on Nov. 30, 2005.[26]

Measurements were taken by the Gas Chromatograph Mass Spectrometer, or GCMS, which identifies different atmospheric constituents by their mass. Analysis of the GCMS findings determined that the methane on Titan was composed of Carbon-13, the isotope of carbon associated with abiotic origins, whereas living organisms have a preference for Carbon-12. Each Carbon-13 atom has an extra neutron in its nucleus, making Carbon-13 atoms slightly heavier than Carbon-12 atoms. NASA scientists examining the carbon isotopes found in the methane

---

[26] Goddard Space Flight Center, "Titan's Mysterious Methane Comes from Inside, Not the Surface," SpaceRef.com, Nov. 30, 2005, at http://www.spaceref.com/news/viewpr.html?pid=18410.

on Titan did not observe the Carbon-12 enrichment that would be suspected if the methane on Titan resulted from organic processes.

Titan has hundreds of times more liquid hydrocarbons than all the known oil and natural gas reserves on Earth, according to a team of Johns Hopkins scientists reporting in Feb. 2008 from data collected by the Cassini-Huygens probe.[27]

"Several hundred lakes or seas have been discovered, of which dozens are estimated to contain more hydrocarbon liquid than the entire known oil and gas reserves on Earth," wrote lead scientist Ralph Lorenz, in the Jan. 29, 2008, issue of the *Geophysical Research Letters*.[28] Lorenz also reported dark dunes running along the equator covering twenty percent of Titan's surface, containing a volume of hydrocarbon material several hundred times larger than Earth's coal reserves.

"Titan is just covered in carbon-bearing material – it's a giant factory of organic chemicals," Lorenz wrote.

The Lost City Hydrothermal Field

Gold began his book, *The Deep, Hot Biosphere*, with a discussion of the deep-sea-diving submarine Alvin's exploration of sea vents along the East Pacific Rise, northeast of the Galapagos Islands. In 2000, the Alvin found a remarkable sub-marine ecosystem in

---

[27]   "Titan's surface organics surpass oil reserves on Earth," European Space Agency (ESA) Space Science, Feb. 13, 2008, at http://www.esa.int/esaSC/SEMCSUUHJCF_index_0.html.

[28]   Ibid.

the mid-Atlantic Ridge, at depths of four to five miles below the surface of the ocean. Termed the "Lost City," this hydrothermal field was living off deep-earth hydrocarbon that was venting out calcium carbonate chimneys that reached up almost 100 yards from the ocean floor.

The scientific exploration of the Lost City proved two things. First, it confirmed Gold's hypothesis that sea-bottom life derives nourishment not from photosynthesis, but from the abiotic hydrocarbons venting from deep within the Earth onto the sea floor. It also backed up the theory that deep-earth, deep-water hydrocarbons are abiotic in nature, formed according to the laws established in the Fischer-Tropsch equations.

In the Feb. 1, 2008, issue of *Science Magazine,* Giora Proskurowski of the School of Oceanography at the University of Washington in Seattle published an article entitled, "Abiogenic Hydrocarbon Production at Lost City Hydrothermal Field."[29] Here, Proskurowski reported on research led by the University of Washington and the Woods Hole Oceanographic Institute that sampled the hydrogen-rich fluids venting from the Lost City's "chimneys." Remarkably, Proskurowski and his team found the hydrogen-rich fluids were produced by the abiotic synthesis of hydrocarbons caused by the simple interaction of

---

[29] Giora Proskurowski, Marvin D. Lilley, Jeffery S. Seewald, Gretchen L. Früh-Green, Eric J. Olson, John E. Lupton, Sean P. Sylva, and Deborah S. Kelley, "Abiogenic Hydrocarbon Production at Lost City Hydrothermal Field," *Science Magazine,* Feb. 1, 2008, Vol. 319, No. 5863, pp. 604-607, at http://www.sciencemag.org/content/319/5863/604.short.

seawater with the rocks underneath the hydrothermal vent field.

"Low-molecular-weight hydrocarbons in natural hydrothermal fluids have been attributed to abiogenic production by Fischer-Tropsch type (FTT) reactions, although clear evidence for such a process has been elusive," Proskurowski and his team wrote in the abstract to the article. "Here, we present concentration, and stable radiocarbon isotope, data from hydrocarbons dissolved in hydrogen-rich fluids venting at the ultramafic-hosted Lost City Hydrothermal Field." Radiocarbon evidence ruled out seawater bicarbonate as the source for the FTT reactions, suggesting that an inorganic carbon source derived from the mantle of the earth was leached from the host rocks. "Our findings illustrate that the abiotic synthesis of hydrocarbons in nature may occur in the presence of ultramafic rocks, water, and moderate amounts of heat."

Ultramafic rocks are igneous and meta-igneous rocks typically found in the earth's mantle. Proskurowski's paper specifically cited the FTT equations in describing how a process called "serpentinization" creates a reducing chemical environment characterized by high hydrogen concentrations suited to abiotic hydrocarbon productions. The serpentinization equations, well understood by scientists since at least 1938, show how the abiotic process works by forming serpentinite from olivine, a magnesium iron silicate found commonly in the earth's mantle. A breakthrough in the FTT equations involved the realization that FTT reactions can occur in deep underwater hydrothermal conditions, where dissolved carbon dioxide is the carbon source used to combine with the hydrogen produced by serpentinization to form the simple C1-C4 hydrocarbon chains the Lost Sea scientists have discovered so far.

Proskurowski ruled out seawater bicarbonate as the carbon source for the observed FTT reactions, insisting that "a mantle-derived inorganic carbon source is leached from the host rocks." Affirming this point, Proskurowski concluded the article by noting, "Hydrocarbon production by FTT could be a common means for producing precursors of life-essential building blocks in ocean-floor environments or wherever warm ultramafic rocks are in contact with water."

Hydrocarbons in Deep Earth

On March 18, 2011, a seminal paper authored by scientists from the University of California at Davis, the Lawrence Livermore National Laboratory, and Shell Products & Technology, entitled, "Stability of hydrocarbons at deep Earth pressures and temperatures," was accepted for publication in the Proceedings of the National Academy of Sciences.[30]

---

[30] Leonardo Spanu, Davide Donaldio, Detlef Hohl, Eric Schwegler, and Guilia Galli, "Stability of hydrocarbons at deep Earth pressures and temperatures," *Proceedings of the National Academy of Sciences of the United States of America,* approved March 18, 2011, at http://www.pnas.org/content/108/17/6843.full.

The importance of the paper was that it revealed how hydrocarbon may be formed from methane deep within the earth at extreme pressures and temperatures. Now, scientists were beginning to establish that higher-chain hydrocarbons were also formed deep within the earth through abiotic processes.

"Our simulation study shows that methane molecules fuse to form larger hydrocarbon molecules when exposed to the very high temperatures and pressures of the Earth's upper mantle," explained UC Davis chemistry and physics professor Giulia Galli, a co-author of the study.[31]

Still, a press release issued jointly by UC Davis and the Lawrence Livermore National Laboratory softened the announcement of the findings on abiotic oil by being sure to mention that, of course, we all know the hydrocarbons that are really important are biological in origin.

"Geologists and geochemists believe that nearly all (more than 99 percent) of the hydrocarbons in commercially produced crude oil and natural gas are formed by the decomposition of the remains of living organisms, which were buried under layers of sediments in the Earth's crust, a region approximately 5-10 miles below the Earth's surface," the press release noted, bowing to political correctness.

---

[31] Anne M. Stark, "Hydrocarbons in the Deep Earth," Press Release, Lawrence Livermore National Laboratory, April 4, 2011, at https://www.llnl.gov/news/newsreleases/2011/Apr/NR-11-04-04.html. The identical Press Release issued by UC Davis can be found here: http://www.ls.ucdavis.edu/mps/news-and-research/hydrocarbons-deep-earth.html.

The Fossil Fuel Paradigm Dies Hard

As physicist Thomas Kuhn pointed out in his 1962 book, *The Structure of Scientific Revolutions*, science advances not by the gradual progress of studies that refine major propositions, but by revolutionary theories that disrupt and ultimately supersede previous, nearly universally accepted scientific hypotheses that are shown to be inadequate in comparison.[32]

According to Kuhn, accepted scientific theories form a "paradigm," defined as a series of propositions that constitute the scientific theory. Ptolemaic astronomy, for instance, was a "paradigm" built around the idea that the sun and planets revolved around the earth. The "Copernican Revolution" replaced Ptolemaic astronomy with the understanding that the earth and other planets revolve around the sun. Paradigm shifts, according to Kuhn, involve revolutions, in which new, competing theories appear first as "heresies." Today's orthodox thinkers might call them "conspiracy theories," which have to fight their way to acceptance, against a legion of established opponents.

Fundamentally, the concept of "fossil fuel" violates the Second Law of Thermodynamics, in which we are given to understand that energy dissipates. As an illustration, consider releasing the neck of a blown-up balloon. The air rushes out.

---

[32] Thomas S. Kuhn, *The Structure of Scientific Revolutions*. (Chicago: University of Chicago Press, 1962). The page numbers cited here come from the University of Chicago Third Edition of the book published as a paperback in 1996.

Forcing the air back into the balloon happens only with a new expenditure of energy. Similarly, organic material at death disintegrates into constituent chemicals.

The Bible teaches, "dust unto dust," to explain what happens to the human body at death; nowhere does the Bible admonish, "dust into oil." We bury dead people in part because decomposing bodies emit a foul odor. No grieving relative ever instructs a funeral director to line a casket with plastic and put a spigot onto it because "Auntie is going to turn into Diesel Fuel No. 2."

Within a few decades, Americans will consider it as ridiculous to contemplate that hydrocarbon fuels were ever called "fossil fuels," as today it is considered ridiculous to imagine that the sun and planets in our solar system revolve around the earth.

3

Hubbert's Peak and the Running-Out-of-Oil Scare

The running-out-of-oil scare is built into the myth that oil is fossil fuel. Almost unconsciously, Americans parrot the conviction oil is fossil fuel, without realizing that by doing so, we are affirming an oil fear developed and fostered by Malthusians for the care and feeding of big oil profits.

The Logic of Oil

If there were only so many ancient forests and dinosaurs available to rot into oil, then there is only so much oil available in the earth. Only a finite number of dinosaurs ever lived, so there's a finite amount of oil. So when we run through all the oil these decaying ancient residues produced, we're done. Built into the logic of the scientific paradigm that has become the "fossil-fuel theory" is the concept that sooner or later we have to run out. In other words, the peak oil theory and the fossil fuel theory are tautologies, two self-reinforcing concepts, neither one of which can be disproven because each implies the other. If oil is fossil fuel, then we are necessarily running out. If we are running out, then oil has to derive from a limited and non-renewable natural resource, such as decaying dinosaurs and ancient forests, or plankton and algae left over from similarly ancient geological periods.

The peak oil theory stands in relation to the biological theory of the origin of oil as a self-evident corollary, for which empirical proof is irrelevant. If peak oil fails to occur at predicted future times, the time prediction for oil depletion can simply be moved to an even more future date. Those who conclude oil is biologic in origin are required to believe, as a matter of logic if not faith, that peak oil followed by oil depletion will happen, if not now, then sooner or later.

The abiotic theory of the origin of oil does not require reaching the conclusion that the world is running out of oil. If the earth produces oil as a natural substance in an on-going manner, then very possibly the earth will never stop producing oil. In other words, we may never run out of oil. Moreover, the abiotic theory implies that oil is a renewable resource.

With the abiotic theory of the origin of oil, we can engage in a scientific calculation to figure out if and when oil depletion will occur. The abiotic oil calculation for oil depletion will depend on estimating current worldwide oil consumption rates, accurately estimating oil reserves, and accurately calculating oil replacement rates. If available oil reserves are dramatically larger than currently estimated and replacement rates more rapid than ever imagined, then worldwide oil consumption may never outpace worldwide oil supply. Here, the assumption oil is abiotic in origin has two advantages over the fossil fuel theory: deep-earth and deep-water abiotic oil reserves may be plentiful across the globe, including at great depths below the oceans, and its production is not limited by the presence of any ancient organic material needed to rot into oil.

The consequences of this debate are both economic and political. If oil is a "fossil fuel" and worldwide oil depletion is inevitable, then industrial society based on the expenditure of hydrocarbon fuel is necessarily threatened unless we develop alternative fuels or conserve oil. This is the calculation politicians opposed to hydrocarbons for ideological reasons use to argue that the development of biofuels, as well as wind and solar energy, are required if we are to maintain our standard of living and achieve continued economic growth. But if geo-scientists have dramatically underestimated the quantity of existing oil reserves by a fossil-fuel bias that never anticipated how abundant deep-earth and deep-water oil actually is or how rapidly the earth continues to produce oil, then oil depletion may not be an imminent reality, regardless of how rapidly the rate of worldwide oil consumption increases. In a world of abundant abiotic oil, alternative energy technologies including biofuels, as well as wind and solar power, could be largely ignored as unnecessary, unless such alternative energy sources prove to be equally robust in their energy output and efficiency as hydrocarbon fuels, and equally or more reasonably priced.

Our so-called "addiction to oil" is only a detriment to global economic advancement if the fossil fuel theory is correct and oil depletion is inevitable. If peak oil concerns turn out to be nothing more than a colossal hoax and oil remains in abundant supply at reasonable prices despite increased worldwide consumption, the global economy can continue its "addiction to oil" without worrying that we are running out of oil. Eliminating the fear that the world is running out of oil eliminates an urgency to experiment with or to implement alternative fuels including biofuels, wind energy, and solar energy as long as these energy technologies remain less energy-efficient, less reliable, and more costly than using oil and natural gas. Simply put, eliminating the fear the world is running out of oil eliminates one of the key ideological underpinnings of the alternative energy movement. This is why those ideologically opposed to using hydrocarbon fuels can be expected to fight against any alternative theory that suggests oil is not a fossil fuel, organic in its origin.

Hubbert's Peak

Probably the most famous formulation of the running-out-of-oil scare is known as "Hubbert's Peak."

In 1956, a geophysicist working in the Shell Oil research lab in Houston, Texas, named M. King Hubbert, published a graph that predicted that US oil production would peak in the

1970s. Hubbert's graph looked like a normal "bell-shaped" distribution curve – in other words, the graph showed almost no production of oil in the early 1900s, then the curve rises to a top point in the early 1970s, from which it drops off gradually until there is no more US oil production at all by the year 2050. In the various accounts of how Hubbert derived his peak graph, there is no indication the diagram resulted from a rigorous scientific examination of available empirical evidence. Instead, anecdotal accounts of how the spark of inspiration hit Hubbert give the impression that he came up with the idea as a thought experiment, almost as if he first formulated the concept by drawing on a napkin at lunch.

Because the graph rises on the page like a mountain, the analysis has become known as "Hubbert's Peak." The name also stuck because "peak" suggests we will reach a high point of oil production from which oil production rates will inevitably drop, ultimately falling back to zero. Hubbard predicted that the United States would hit peak oil production in the 1970s. Today, as US oil production continues to increase, Hubbert's initial prediction appears to have been somewhat hysterical. Instead of abandoning the peak oil theory altogether, petro-scientists in the 1990s merely revised their calculations to predict the world would hit peak oil production somewhere between 2004 and 2008. This alone should have disqualified the Hubbert's Peak as scientific theory. To be scientifically rigorous, a theory must be subject to refutation if empirical results turn out differently than predicted. A theory that can neither be validated nor invalidated by empirical results is more properly considered a prejudice than a rigorously postulated scientific hypothesis.

Princeton Professor Emeritus Kenneth S. Deffeyes, a geologist who worked with M. King Hubbard at Shell Oil in the 1950s, noted that Hubbert made his 1956 prediction at a meeting of the American Petroleum Institute in San Antonio. Deffeyes relates that the Shell Oil head office was on the phone with Hubbert right down to the last five minutes before his talk, asking Hubbert to withdraw his prediction. Deffeyes commented that Hubbert had "an exceedingly combative personality," and he went ahead with the announcement, despite Shell Oil reservations.

"I went to work in 1958 at the Shell research lab in Houston, where Hubbert was the star of the show," Deffeyes wrote. "He had extensive scientific accomplishments in addition to his oil prediction. His belligerence during technical arguments gave rise to a saying around the lab: "That Hubbert is a bastard, but at least he's our bastard."[33] That Deffeyes felt it necessary to make a point of crediting Hubbert with scientific accomplishments in addition to his oil prediction hypothesis almost sounds like an excuse. Perhaps Deffeyes was stretching to assert Hubbert's Peak was scientifically derived, not simply

[33] Kenneth S. Deffeyes, *Hubbert's Peak: The Impending World Oil Shortage* (Princeton, New Jersey: Princeton University Press, 2001), pp. 1-3.

the idle speculation of an oil company executive, based on little more than a fuzzy intuition of how the bell-shaped curve every college student learns in Statistics 101 might be applied to shed light on a question as complex as estimating and predicting reliably future oil depletion rates, both in the United States and ultimately on a global basis.

Gloom, Doom, and the Psychology of High Priced Oil

Truthfully, major US oil companies embraced Hubbert's Peak almost immediately. If oil was running out, it would eventually become scarce, and scarce resources can justifiably command premium prices. US oil companies had no economic reason to alert the American population to the reality that oil is not fossil fuel, even when abundant deep-earth and deep-water oil was being found and brought to production at affordable prices, at depths where no land animal, forest, plankton, or algae ever lived, not now or in ancient times. But those whose thinking is locked in Hubbert's Peak predictions are loathe to abandon the theory. Not unexpectedly, Deffeyes opened up his 2001 book, entitled *Hubbert's Peak: The Impending World Oil Shortage*, with the following paragraph on the opening page of his "Overview":

> *Global oil production will probably reach a peak sometime this decade. After the peak, the world's production of crude oil will fall, never to rise again. The world will not run out of energy, but developing*

alternative energy sources on a large scale will take at least 10 years. The slowdown in oil production may already be beginning; the current price fluctuations for crude oil and natural gas may be the preamble to a major crisis.[34]

In 2003, Princeton University Press published the sixth printing of Deffeyes' book, issued in paperback. In this "revised and updated" paperback, the chart presented on page 3 as "Hubbert's original 1956 graph," had to be modified to change Hubbert's original prediction that US oil production would peak in the early 1970s. As we will see in a few pages, US Department of Energy statistics show US oil production continues to increase, not decline as Hubbert's Peak advocates have predicted. Deffeyes in the 2003 revision of his book merely added additional curves to move to a later date the anticipated oil depletion date. The corrected chart shows actual US oil production for 1956 through 2000 at much higher levels than Hubbert originally predicted. "Since 1985, the United States has produced slightly more oil that Hubbert's prediction," Deffeyes conceded, "largely because of successes in Alaska and in the far off-shore Gulf Coast." The point is that instead of conceding that empirical data proved Hubbert's hypothesis faulty, Deffeyes altered the predictions to preserve the theory, despite empirical data to the contrary. In doing so, Deffeyes affirmed Hubbert's Peak is properly considered as either a tautological restatement of

---

[34] Ibid., p. 1.

the fossil fuel theory itself or a matter of faith drawn from the realm of near-religious beliefs to be stated as if it were a scientific hypothesis.

Now-deceased Houston investment banker Matthew R. Simmons, a life-long proponent of peak oil, published in 2005 his seminal book, Twilight in the Desert: The Coming Saudi Oil Shock and the World Economy, articulating his theory that oil depletion was occurring in Saudi Arabia, the world's leading country in oil production. Simmons declared that following a remarkable string of exploration successes from 1940 through 1968, oil producing giant Saudi Arabia had hit a brick wall. "For the next three decades, Saudi Aramco employed the best exploration technologies anywhere available to bulk up its portfolio of world-class oilfields," Simmons wrote. "As with exploration elsewhere around the world, the effort became a high-stakes game requiring substantial risk for elusive rewards." For Saudi Arabia, Simmons concluded, exploration for new oil reserves since 1968 has produced "very meager payoffs."[35] A believer in the fossil fuel theory, Simmons concluded Saudi Arabia faced an inevitable dimming future of its oil industry, playing out a script that "was written in the geology eons ago."[36] Simmons came to see Saudi Arabia as living off production in aging super-fields, unable to find additional giant or super-giant oil fields. He concluded that twilight was descending over the oil fields of Saudi

---

[35] Matthew R. Simmons, *Twilight in the Desert: The Coming Saudi Oil Shock and the World Economy* (Hoboken, N.J.: John Wiley & Sons, Inc., 2005), p. 231.

[36] Ibid., p. 281.

Arabia. If Saudi Arabia was facing oil depletion, the implication was that it was only a matter of time before oil fields around the world would face oil depletion, making the future of oil suspect and problematic in a world economy dependent upon hydrocarbon fuels.

A Malthusian Future?

Even today, Hubbert's Peak remains an almost universally accepted axiom among petroleum geologists, accepted as a matter of faith as being a true and established law. Usually, after stating as a truism that the United States has no choice but to increase our dependency on foreign oil because we have already hit peak oil, traditional thinkers writing about energy lament that in a relatively short period of time, such as perhaps in the next 100 years, the world is going to use up what took nature millions of years to create. Consider this statement in the thirtieth-year update of the famous 1972 MIT study entitled *Limits to Growth*, an early statement of "sustainable growth" theories that expresses at its heart a typically pessimistic evaluation that world economies are exhuasting available natural resources:

> Optimists and pessimists differ by a few decades in the timing of its [oil's] production peak. But there is substantial consensus that petroleum is the most limited of the important fossil fuels, and its global pro-

duction will reach a maximum sometime during the first half of this century.[37]

Simply put, politically correct dogma typically considers oil to be a non-renewable energy source. Moreover, industrial societies such as ours are blamed for engaging in an irresponsible burning of fossil fuels.

Consider this dire warning from an analyst who is further convinced our burning of oil contributes to global warming:

> *Nature took about a million years to lay down the amount of fossil fuel that we now burn worldwide every year – and in doing so it seems that we are causing rapid change of the Earth's climate. Such a level of exploitation is clearly not in balance, not harmonious and not sustainable.*[38]

Authors believing hydrocarbon fuels are fossil-produced have no choice but to issue public policy suggesting Americans must wean themselves off oil. The only alternatives such traditional thinkers can envision involve scaling back the US economy and/or dramatically restricting our lifestyles, while demanding new legislation that mandates the use of alternative fuels, including both solar and wind, whether or not the alter-

---

[37] Donella Meadows, Jorgen Randers, and Dennis Meadows, *Limits to Growth: The 30-Year Update* (White River Junction, Vermont: Chelsea Green Publishing Company, 2004), p. 87.

[38] John Houghton, *Global Warming: The Complete Briefing* (Cambridge, England: Cambridge University Press, Third Edition, 2004), p. 199.

natives are robust or affordable. Doom-and-gloom is the inevitable church hymn refrain of those who choose to believe with near religious fervor the fossil fuel theory of the origin of oil.

Reading book after book predicting a gloom-and-doom energy future, we are left with the conclusion that fossil fuel advocates are locked into the type of thinking best characterized by Thomas Malthus. In his famous 1789 essay, Malthus predicted that population would ultimately outstrip our ability to produce food, resulting in a series of crises such as war and famine, which in turn would cut back populations to more manageable levels. Malthus proposed this as a mathematical law that governed and restricted population growth. Since population growth proceeds at a geometric rate (i.e., 2, 4, 8, 16, etc.) and food production proceeds at an arithmetic rate (i.e., 1, 2, 3, 4, 5, etc.) there was no way the success of population growth could not result ultimately in disaster to those very populations which had managed so successfully to grow.

Malthus is famous not because his theory was right, but because human experience has proven him wrong. Malthus failed to anticipate adequately the human genius for adaptation, invention, and technological advancement. Populations have grown widely beyond all the limits Malthus thought possible. Yet, even today, with a world population measured in billions that Malthus never imagined possible, Malthusians will insist Malthus was right. Doom-and-gloom Malthusians continue to theorize that worldwide famine is inevitable, even if they follow the convenient logic of peak oil theorists as they continue to push back the date predicting when the world calamity will occur.

More Worldwide Oil Reserves Today than Ever

The problem with this the doomsday analysis that we are running out of oil is that worldwide the available data does not conform to the reality predicted. Government and industry-collected statistics prove that worldwide we are now sitting on more proven petroleum reserves than ever before, despite the increasing rate at which we are consuming petroleum products worldwide.

Moreover, new and gigantic oil fields are being discovered at an increasing rate, at deep-earth and deep-water levels, at depths below the surface of the earth that the fossil fuel theory would never have imagined as possible.

Let's examine the evidence and see if it doesn't sound a lot more like the abiotic theory is the appropriate model for comprehending how the earth produces hydrocarbon fuels naturally.

According to the Energy Information Administration of the US Department of Energy, there are more proven crude oil reserves worldwide than ever in recorded history, despite the fact that worldwide consumption of crude oil has doubled since the 1970s. The EIA reports that in 2009, worldwide crude oil reserves exceeded 1.34 trillion barrels, compared with

1.02 trillion barrels in 2000. This is a long-term trend upward. In 1980, the EIA estimated worldwide proven oil reserves at 645 billion barrels; in 1985, 700 billion barrels; in 1990, 1 trillion barrels; in 1995, 999 billion barrels; in 2000, 1.02 trillion barrels; and in 2005, 1.28 trillion barrels. This data represents a nearly unbroken progression of increasing proven oil reserves over the last quarter century – hardly the pattern we would expect to find if the world were really running out of oil.[39] The truth is the world did not hit peak oil production in the 1970s, as Hubbert's Peak initially predicted, or at any of the other dates the Hubbert's Peak revised model postulated. Looking at the data trend, peak oil is nowhere in sight, not when over the past quarter-century authoritative estimates of proven oil reserves worldwide only continue to increase.

A further indication that peak oil theory is a hoax occurred in July 2008, when oil prices spiked to an all-time high of $147 a barrel, only to recede to under $40 a barrel before the end of the year. When oil prices spiked, peak oil theorists claimed the dramatic price increase was proof oil production rates had slowed to create disequilibrium with increasing world oil demand.

The truth was that oil prices are largely determined by supply and demand. Oil traders, including those specula-

---

[39] The data are taken directly off spreadsheets published by the Energy Information Administration, US Department of Energy. The data represent the official energy statistics from the US government. See: http://www.eia.gov/emeu/international/oilreserves.html.

tors bidding in the oil futures markets, had not realized until after the bank crisis of July 2008 that worldwide oil demand was likely to decrease dramatically. In July 2008, oil speculators had underestimated the magnitude of the worldwide economic recession likely to result from the bursting of the US mortgage bubble. By the end of 2008, even oil speculators realized oil demand had declined dramatically worldwide, with the result that oil prices receded rapidly from their $147 a barrel high.

Yet even in July 2008, with oil at $147 a barrel, there was no shortage of oil in the United States – no rationing or gas lines at service stations was required. In other words, the all-time high price of oil in July 2008 was not proof that oil had become inherently scarce or in irreversible short supply.

Even those predisposed to view peak oil theory favorably, such as ecologist George Wuerthner, have to admit problems with the concept that the maximum worldwide oil production rate has been or will soon be reached. "By 2000, the point when Hubbert estimated that we would reach global Peak Oil we would have only around 625 billion barrels of oil left," Wuerthner wrote in *Counterpunch*, on March 29, 2012.

"Just the 558 billion barrels of proven reserves known to exist in Saudia Arabia and Venezuela alone (and a lot more in-place resources) is nearly equal the total global oil supplies that Hubbert estimated would remain in global reserves," he continued. "Obviously, Hubbert's global estimates were way too low.

The world has already burned through more than a trillion barrels of oil, clearly demonstrating how far off his predictions of oil supplies were. The estimated 'proven reserves' left globally are today more than 1.3 trillion barrels from the top seventeen oil producing countries alone."[40]

[40] George Wuerthner, "The Myth of Peak Oil," *Counterpunch*, March 29, 2012, at http://www.counterpunch.org/2012/03/29/the-myth-of-peak-oil/print.

4

Deep-Earth and Deep-Water Oil

Nothing has challenged the fossil fuel theory more than the advances made in deep-earth and deep-water drilling, the fastest-growing segment of the energy industry in the last twenty years.

Only two years after an explosion on the British Petroleum Deepwater Horizon oil platform in the Gulf of Mexico, international companies are ready to expand deep-water in Mexican and Cuban waters beyond US control, while new deep-water drilling is scheduled off the coast of East Africa and in the Mediterranean. Despite the moratorium the Obama administration placed on deep-water exploration and production in the Gulf after the Deepwater Horizon disaster, by early 2012 forty rigs were drilling in the Gulf, compared to only twenty-five a year earlier. In early 2012, British Petroleum had five rigs drilling in the Gulf, the same number as before the disaster, and the company has plans to add three more rigs in the Gulf before the end of the year. The Energy Information Administration expects oil production in the Gulf will expand from its level of 1.3 million barrels a day, one-quarter of total US domestic oil production, to 2 million barrels a day by 2020.[41] Without the resraints placed on Gulf drilling by the Obama administration the oil production numbers would be dramatically higher.

Deep-water drilling typically involves offshore rigs drilling on the continental shelves around the world in water thousands of feet in depth. The advantage of offshore drilling is that oil-rigs get free passage through water before deep-earth drilling begins, allowing deeper drilling into the earth with less technical difficulty and expense. As recently as twenty years ago, oil drilling technology had not advanced to the point where drilling at these depths was feasible, either technically or economically. Within the next twenty years, technological advances will probably make drilling even more feasible and economical in deeper waters off the continental shelf, where abiotic oil theory predicts that abundant new reserves unknown today should be found.

---

[41] Clifford Krauss and John M. Broder, "Deepwater Oil Drilling Picks Up Again as BP Disaster Fades," New York Times, March 4, 2012, at http://www.nytimes.com/2012/03/05/business/deepwater-oil-drilling-accelerates-as-bp-disaster-fades.html?pagewanted=all.

Mexico: The Cantarell Oil Field

The Cantarell oil field, discovered in 1976, and supposedly named after the fisherman who reported an oil seep to Mexican government authorities in the Campeche Bay, has largely been responsible for keeping Mexico in the top ten oil producing countries in the world.

In the 1970s, geophysicist Glen Penfield established that a massive meteor hit the earth at the end of the Cretaceous Period, approximately sixty-five million years ago, in the Yucatàn near the town of Chicxulub. In the 1980s, physicist Luis Alvarez and his geologist son Walter had suggested in their independent studies that an impact meteor hitting earth between the Cretaceous and Tertiary Periods, at the end of the Mesozoic Era, caused the extinction of the dinosaurs. Whether the Chicxulub meteor was the culprit that killed the dinosaurs remains debatable. What appears more likely is that the impact sufficiently fractured the Gulf of Mexico bedrock off the Yucatàn coast, creating the Cantarell oil field and opening up oil exploration opportunities throughout the Gulf.

The Chicxulub impact crater is enormous, estimated to be 100 to 150 miles (160 to 240 kilometers) wide. The seismic shock of the meteor deeply fractured the bedrock below the Gulf, and set off a series of tsunami waves that caused a huge section of land to break off and fall back into the crater under water. The severe fracturing of the bedrock in the Gulf facilitated the flow of liquids and gases from the deep earth below.[42]

Until the 1960s, geologists considered collisions of extra-terrestrial objects with the earth interesting, but not necessarily important. Since Cantarell was discovered, geologists have come to realize that the intense shock waves generated in meteor impact events have significantly shaped the earth's surface, distributed its crust, and fractured its bedrock. Over 150 significant geological structures, many masked by subsequent sedimentary deposits, have been identified worldwide, ranging from circular impact bowls measuring from only a few kilometers in diameter to as many as 200 kilometers (approximately 125 miles) in diameter. Cantarell has stimulated interest in meteor impact structures as locations to explore as potential oil-producing sites.[43]

Since 2005, Petroleos de Mexico, known as "Pemex," Mexico's state-owned oil company, has discovered two deep-water oil fields off the shores of Veracruz, in what is known as Coatzacoalcos Profundo in the Gulf; Noxol-1, located 63 miles northwest of Coatzacoalcos off the coast of Veracruz, situated in 935 meters (3,067 feet) of water; and Lakach-1, located 81 miles northwest of Coatzacoalcos, situated in waters measuring 988 meters in

[42] Alan R. Hildebrand, Glen T. Penfield, David A. Kring, Mark Pilkington, Antonio Carmargo Z., Stein B. Jacobsen, and William V. Boynton, "Chicxulub Crater: A possible Cretaceous/Tertiary boundary impact crater on the Yucatàn Peninsula, Mexico," *Geology*, September 1991, Vol. 19, No. 9, p. 867-871.

[43] Richard R. Donofrio, "North American impact structures hold giant field potential," Oil & Gas Journal, May 11, 1998, at http://parwestlandexploration.com/docs/og.pdf.

depth (3,241 feet). These two discoveries have caused Pemex to estimate that Coatzacoalcos Profundo contains reserves amounting to ten billion barrels of oil.[44] Additionally, two years after the Deepwater Horizon catastrophe, Pemex has announced plans to deploy two state-of-the-art drilling platforms just south of the maritime boundary with the United States, with one rig drilling in 9,514 feet of water, and the other rig drilling in 8,316 feet of water.[45] By definition, ultra-deepwater drilling involves anything more than 5,000 feet of water. Deepwater Horizon, when it exploded, was in about 5,100 feet of water. Since Penemax decided in 2004 to expand drilling beyond the shallow waters close to the coast, the government-owned company has drilled 16 wells at increasing depths, with two in ultra-deepwaters, Caxa and Kunah, with the latter at 6,500 feet.[46]

Saudi Arabia: Basement Tectonics

The reason Saudi Arabia has abundant oil but neighboring countries such as Afghanistan do not, is not that the dinosaurs bypassed Afghanistan to herd in Saudia Arabia and die in a big heap at the end of the Mesozoic Era. The reason is that the bedrock beneath the Saudi oil fields is deeply fractured.

---

[44] "Coatzacoalcos Profundo," in Offshore Field Development Projects, SubseaIQ.com, last updated March 11, 2008, at http://www.subseaiq.com/data/Project.aspx?project_id=493&AspxAutoDetectCookieSupport=1.

[45] Tim Johnson, McClatchy Newspapers, "Pemex set to drill ultradeep oil wells in Gulf of Mexico, April 8, 2012, at http://www.sacbee.com/2012/04/08/4398071/pemex-set-to-drill-two-ultradeep.html.

[46] Ibid.

An important, but largely neglected study of the bedrock underlying the Saudi oil fields provides strong evidence that the Saudi oil fields resulted from fractures and faults in the basement rock. The study, "Basement Tectonics of Saudi Arabia as Related to Oil Field Structures," was published in 1992 by H.S. Edgell, a geologist at the King Faud University of Petroleum & Minerals, in Dhahran, Saudia Arabia. Edgell argued that the Saudi oil fields, including the giant field at Ghawar, were produced by oil produced in the mantle of the earth seeping up through bedrock fractures that lie beneath the oil fields.[47]

"All the oil fields of Saudia Arabia are of the structural type and they all lie in the northeastern part of the country, including the Saudi offshore portion of the Persian Gulf," Edgell wrote. "These oil field structures are mostly produced by extensional block faulting in the crystalline Precambrian basement along the predominantly N-S Arabian Trend which constitutes the 'old grain' of Arabia."[48] The Precambrian period dates back geologically some 4.6 billion years, to the origin of the earth, until some 570 million years ago. Dinosaurs did not appear on the earth until much later, during the Mesozoic Era, beginning 250 million years ago, a considerable distance in time from the Precambrian Era.

---

[47] H. S. Edgell, "Basement Tectonics of Saudi Arabia as Related to Oil Field Structures," in M.J. Rickard, H.J. Harrington, and P.R. Williams (editors), *Basement Techtonics 9: Australia and Other Regions,* Proceedings of the Ninth International Conference on Basement Techtonics, held in Canberra, Australia, July 1990 (Dordrecht/Boston/London: Kluwer Academic Publishers, 1992), pp. 169-193.

[48] Ibid., p. 169.

Edgell's study would argue that oil in Saudi Arabia is abundant because fault patterns in the underlying bedrock permit oil from the earth's mantle to seep upward, into the many porous sedimentary strata lying above. Edgell is not shy about advancing this conclusion: "All the known oil fields of Saudi Arabia and its offshore are thus related to four major directions of basement faulting, namely N-S, NE-SW, NW-SE, and E-W."[49] And again:

> Anticlinal or domal structures in the sedimentary sequence of the northeastern Arabian Platform and its offshore extension contain all the known oil and gas fields of Saudi Arabia. These currently comprise some fifty-six oil fields, all of which owe their origin to deep-seated tectonic movements in the Precambrian crystalline basement.[50]

Translated into simple terms, Edgell is telling us to forget about dinosaurs, ancient forests, plankton, and algae. Saudi Arabia has abundant oil because the fault pattern under Saudi Arabia permits oil from the earth's mantle to flow upward.

As noted earlier, Matthew Simmons, in his book entitled *Twilight in the Desert,* painted a grim picture of Saudi Arabian oil prospects, arguing that even the giant oil field of Ghawar is depleting and is increasingly cut by water to increase production.

---

[49]   Ibid.

[50]   Ibid., p. 170.

Simmons argued that Aramco is going after the "last of the easily produced, free-flowing oil in the most prolific parts of Ghawar."[51]

Simmons' dire predictions stand in direct contrast to the Saudi's much more optimistic view. Saudi Minister of Petroleum and Mineral Resources Al-Naimi, at a conference on Saudi oil held in Washington, D.C., in April 2004, said that Saudi oil reserves have been dramatically underestimated:

> Saudi Arabia now has 1.2 trillion barrels of estimated reserve. This estimate is very conservative. Our analysis gives us reason to be very optimistic. We are continuing to discover new resources, and we are using new technologies to extract even more oil from existing reserves.[52]

In 2011, the Energy Information Administration estimated Saudi oil resources at approximately 261.9 billion barrels, one-fifth of the world's proven oil reserves, but only 20 percent of Al-Naimi's 2004 estimate.[53] Even Simmons had to acknowledge how difficult it is to obtain accurate data on

---

[51] Matthew R. Simmons, *Twilight in the Desert*, op.cit., page 170.

[52] Tim Kennedy, "Saudi Oil Is Secure and Plentiful, Say Officials," Arab News, April 29, 2004, at http://archive.arabnews.com/?page=6&section=0&article=44011&d=29&m=4&y=2004.

[53] Energy Information Administration, US Department of Energy, "Saudi Arabia, Country Profile," last updated January 2011, at http://205.254.135.24/emeu/cabs/Saudi_Arabia/Full.html.

Ghawar, the Saudis' largest field, or on any specific details of Saudi production.

> *Ghawar is well known as the world's largest oilfield within the petroleum industry and among analysts and energy journalists. But few people, even among the world's more knowledgeable energy experts, know anything more about Ghawar beyond its colossal size. Rarely has any data been published that provided details about the performance and parameters of this greatest of all oilfields.*[54]

How then could Matthew Simmons be confident Saudi Arabian oil had reached peak production, if there is no reliable estimate of the total amount of oil reserves Saudi Arabia possesses? If the Saudis have benefited from basement tectonics that allow deep-earth oil formed in the mantle of the earth to freely flow upward, how possibly can anyone estimate the amount of oil Saudi Arabia might have at levels far below the earth's surface? If oil production is abiotic and on-going, we must estimate the rate at which oil is produced within the mantle of the earth, in order to determine if replenishment rates exceed production rates.

---

[54] Matthew Simmons, op. cit., page 155.

Deepwater Horizon Disaster

The explosion on April 20, 2010, on the Deepwater Horizon oil rig, operated in the Tiber Field in the Gulf of Mexico by British Petroleum (BP), caused the deaths of eleven workers and was the largest oil spill in history. After a series of attempts to plug the leak, BP successfully capped the well, stopping the flow of oil into the Gulf for the first time, eighty-six days after the explosion.

In Sept. 2009, at the time the Tiber discovery was announced, Daren Beaudo, a BP spokesman, told the *Washington Post* the discovery rivaled another giant field operated by BP in the Gulf, known as Thunder Horse, then producing as much as 300,000 barrels of oil a day.[55] The communications director for Transocean, the offshore drilling company that owned the Deepwater Horizon rig, announced that the well would be dug in 4,130 feet of water, and drilled another 30,923 feet below the sea floor, with the result that the oil would be brought up to the rig from more than six miles below the surface of the Gulf.

In the feature story on the BP discovery printed on Sept. 3, 2009, the *Washington Post* quoted oil historian Daniel Yergin, chairman of Cambridge Energy Research Associates, as saying the discovery "demonstrates how technology continues to expand the horizon of the Gulf of Mexico." BP said the well struck oil "in

---

55 Steven Mufson, "BP Finds 'Giant' Oil Source Deep Under Gulf of Mexico, *Washington Post*, Sept. 3, 2009, at http://www.washingtonpost.com/wp-dyn/ content/article/2009/09/02/AR2009090203560.html.

multiple reservoirs" in the Lower Tertiary geological zone, a layer of the earth's crust dating back thirty-eight to sixty-eight million years. The newspaper noted more than ten discoveries had been made at that level in the Gulf, including BP's Kaskida find that had estimated reserves of four to six billion barrels. "We view the Lower Tertiary as being one of the next big waves of development in the Gulf of Mexico," Beaudo told the *Washington Post.*

Reporting on Jan. 6, 2011, the National Oil Spill Commission attributed the disaster to a failure of BP management to appropriately evaluate risk factors and to implement the necessary technical and operating safeguards.[56] *The Guardian* in the UK, however, suggested the real culprit was peak oil. "The only long-term answer is to wean ourselves off oil before the post-peak trouble really starts," an environmental blog in the London newspaper proclaimed the day the presidential oil spill commission announced its findings. "It's amazing stuff: energy-dense and easily transported. But alternatives exist, from electric vehicles to biofuels to fuels generated from sunlight. These need investment, but would we really rather spend billions on clean-up operations and lawyers, I hope not."[57]

---

[56] Press Release, "Commission Releases Chapter on BP Well Blowout Investigation in Advance of Full Report," National Commission on the BP Deepwater Horizon Oil Spill and Offshore Drilling, Jan. 6, 2011, at http://www.oilspillcommission.gov/sites/default/files/documents/Advance%20Chapter%20on%20BP%20Well%20Blowout%20Investigation%20Released.pdf.

[57] Damian Carrrington's Environmental Blog, "Deepwater Horizon oil spill: The real cause is peak oil," *Guardian,* Jan. 6, 2011, at http://www.guardian.co.uk/

Abiotic oil observers had a different analysis. The force of the oil flow after the explosion suggests the oil reserves found by the Deepwater Horizon rig had to be enormous, perhaps generating more pressure than current technology could safely handle. Granted, the various studies and legal challenges following the disaster found many operating conditions that could be attributed to negligence. The only solution was to cap the well, at least for now. Most likely, BP will reopen the Tiber well at some unspecified future date, when deep-water and deep-earth technology has further advanced to take into account the pressures and temperatures that will have to be managed if oil and natural gas are to be commercially produced in a safe and economic manner at depths miles down from the surface of the earth.

Brazil: Finally Independent from Biofuels

Petrobras, Brazil's semi-public, partially government-owned oil company is moving Brazil from being nearly 100 percent dependent on foreign oil imports only some fifty years ago, toward becoming a net oil exporter in the next few years. How? Brazil has realized spectacular results by developing the technology to drill ultra-deep offshore wells in Brazil's Barracuda and Caratingua oil fields, in the Campos Basin some fifty miles into the Atlantic Ocean east of Rio de Janeiro. In the process,

environment/damian-carrington-blog/2011/jan/06/bp-oil-spill-peak-oil.

Brazil has rapidly weaned itself off sugar-produced ethanol, once the only fuel produced in Brazil.

According to the Energy Information Administration, Brazil has gone from almost no oil production in 1980 to producing approximately 2.1 million barrels of crude oil a day in 2011. Brazil's oil production has grown at a rate of about 9 percent per year since 1980. The EIA further estimates that Brazil has fourteen billion barrels of proven oil reserves in 2012, the second largest in South America after Venezuela. "Increased domestic oil production has been a long-term goal of the Brazilian government, and recent discoveries of large offshore, pre-salt oil deposits could transform Brazil into one of the largest oil producers in the world." With the country consuming 2.2 million barrels per day, Brazil is about to become oil independent. The EIA has forecast that by 2013, Brazilian oil production would reach three million barrels a day. By the end of this decade, Brazil expects to become a net oil exporter. Brazil's offshore drilling success represents a complete turnaround; in 1953, Brazil's domestic oil production filled only 3 percent of domestic demand.[58]

To develop the oil resources of the Campos Basin, Petrobras formed the Barracuda & Caratingua Leasing Company B.V. (BCLC) as a special purpose corporation established in the Netherlands. In Dec. 2004, BCLC finalized an $2.5 billion agree-

---

[58] "Brazil: Country Analysis," , US Department of Energy, last updated Feb. 28, 2012, at http://www.eia.gov/countries/cab.cfm?fips=BR.

ment with Halliburton's Kellogg Brown & Root subsidiary (KBR) awarding KBR a full engineering, procurement, installation and construction (EPIC) contract for fifty-five offshore wells in the two oil fields (twenty-two horizontal producers and two multi-lateral horizontal producers, as well as eight horizontal injectors and eight piggyback injectors). The contract also specified the construction and installation of two floating, production-storage, offloading vessels (FPSOs). The Barracuda and Caratingua fields are expected to add 30 percent to the current one million barrels per day production rate from the Campos Basin region. The two fields cover a combined area of 230 square kilometers (approximately 145 square miles). The Barracuda and Caratin-gua proven oil reserves are estimated at 1.229 billion barrels. Together they are expected to produce 773 million barrels of oil by 2025.[59]

None of this will impress "peak production" or "fossil fuel" theorists, who expectedly will argue that the Brazil's offshore oil fields, regardless of how large they might be, are doomed to deplete sooner or later. Petrobras has a different vision. If the Campos Field has yielded massive oil deposits, are there other fields on Brazil's intercontinental shelf that remain to be discovered?

In 2007, Brazil announced the discovery of a second huge offshore oil field in the Santos Basin off Brazil's shore, south

---

[59] "Barracuda and Caratingua Fields, Brazil," OffshoreTechnology.com, no date, at http://www.offshore-technology.com/projects/barracuda-caratingua-fields-brazil/.

of the Campos Basin. The Tupi Field in the Santos Basin is estimated to contain between five and eight billion barrels of oil, enough to expand Brazil's 14.4 billion barrels of proven oil reserves by 40 to 50 percent. The ultra-deepwater Tupi field was found under 7,060 feet of water (1.34 miles down), beneath 10,000 feet of sand and rocks (another 1.89 miles down), and finally another 6,600 feet of salt (1.25 miles), for a total of 4.48 miles below the surface of the Atlantic Ocean. In April 2012, Petrobras announced the discovery of the Lula field in the Santos Basin, another in a series of discoveries that could make the Santos Basin equally as productive for Petrobras as the Campos Basis has been.

"Deep" Gas Wells below 15,000 Feet

The Energy Information Administration estimates that world consumption of natural gas is expected to increase from 111 trillion cubic feet in 2008 to 169 trillion cubic feet in 2035.[60] The International Energy Administration's *World Energy Outlook 2011* posed the question: "Are we entering a golden age of gas?"[61] The IEA estimated that conventional recoverable resources of natural gas are equivalent to more than 120 years of current global consumption, while total recoverable re-

---

[60]   US Department of Energy, "International Energy Outlook 2011," Sept. 19, 2011, at http://www.eia.gov/forecasts/ieo/more_highlights.cfm#world.

[61]   International Energy Agency, *World Energy Outlook 2011* (Paris, France: International Energy Agency, OECD, 2011), at http://www.iea.org/weo/docs/weo2011/WEO2011_GoldenAgeofGasReport.pdf.

sources could sustain today's production for over 250 years. Contrary to the expectations of peak production theorists, natural gas resources are abundant in the United States, especially at deep-earth levels.

A "deep" gas well is typically defined as any that produces from a depth below 15,000 feet (2.84 miles). According to the Potential Gas Committee's 2003 Report, there were over 2,500 active natural gas wells producing at or below that level in the United States, drawing from 183 natural reservoirs located primarily in the on-shore and off-shore basins of the Texas and Louisiana Gulf Courses.[62] Today there are some 400,000 producing natural gas wells in the United States; however, few are "deep" gas wells. The US Department of Energy notes that more than 70 percent of the natural gas produced in the United States comes from wells at 5,000 feet or deeper, while only 7 percent comes from formations below 15,000 feet. Yet, at depths below 15,000, the Department of Energy estimates that 125 trillion cubic feet of natural gas is thought to be trapped.[63]

The western world's record for deep-well natural gas exploration and production is held by the GHK Company in Oklahoma. From 1972 through 1974, the company engineered and

---

[62] The National Energy Technology Laboratory (NETL), US Department of Energy, "Exploration & Production Technologies: Advanced Drilling – Deep Trek – Deep Gas Wells," NETL.DOE.gov, at http://www.netl.doe.gov/technologies/oil-gas/EP_Technologies/AdvancedDrilling/DeepTrek/GasWells.html.

[63] US Department of Energy, "'Deep Trek' and Other Drilling R&D," Fossil.energy.gov, at http://fossil.energy.gov/programs/oilgas/drilling/index.html.

drilled two Oklahoma natural gas commercial wells at depths greater than 30,000 feet (approximately 5.7 miles), the #1-27 Bertha Rogers well (total depth 31,441 feet) and the #1-28 E.R. Baden well (total depth at 30,500 feet),[64] both located in the deep and high pressure Anadarko Basin that covers some 50,000 square miles across west-central Oklahoma, the upper Texas panhandle, southwestern Kansas, and southeastern Colorado. Since the company's founding in the mid-1980s, GHK reports drilling and operating 193 wells, the majority of which are below 15,000 feet, all without experiencing a blowout. GHK's success ratio for all drilling operations, including wildcat exploratory drilling, from 1995 to 2005, has been 82 percent.

A study conducted by Mark Snead, Ph.D., the director of the Center for Applied Economic Research at the University of Oklahoma's Spears School of Business, documents the commercially successful deep-well drilling for natural gas in the state:

> Oklahoma has long played an important role in the development of deep drilling. The first hole drilled below 30,000 feet for commercial production purposes was completed in Beckham County in 1972.

---

[64] The GHK Companies, "Anadarko Basin – 1960's through 1980's," GHKco.com, at http://www.ghkco.com/our_operations/; and Oklahoma Historical Society's Encyclopedia of Oklahoma History & Culture, "Andarko Basin," at http://digital.library.okstate.edu/encyclopedia/entries/A/AN003.html.

And continuing:

> *The Anadarko Basin has historically been one of the most prolific natural gas producing regions in the United States and is the location of most of the deep wells in Oklahoma. According to the US Geological Survey, 20 percent of the holes drilled deeper than 15,000 feet prior to 1991 are located in the Anadarko Basin, exceeding the number of deep wells in all drilling regions in the US other than the Gulf of Mexico in the period. Through 1998, 19 of the 52 existing ultra deep wells below 25,000 feet were drilled in the Anadarko Basin.*

> *Through 2002, the Potential Gas Committee reports that a total of 1,221 producing deep wells were completed in Oklahoma at an average depth of 17,584 feet, with 775 of these wells currently active.*[65]

The success with deep drilling of natural gas resources has been experienced across the United States:

> *The overall success rate of deep wells has been remarkably good. In a sample of 20,715 deep wells drilled in*

---

[65] Mark C. Snead, Ph.D., Director, Center for Applied Economic Resarch, Spears School of Business, "The Economics of Deep Drilling in Oklahoma," Oklahoma State University, Stillwater, Oklahoma, February 2005, at http://economy.okstate. edu/caer/files/economics_of_deep_drilling.pdf.

> *the US through December 1998, 11,522 (56 percent)*
> *are classified as producing gas and/or oil wells, with*
> *gas wells comprising nearly 75 percent of producing*
> *wells. Of the 1,676 wells exceeding 20,000 feet, 974*
> *(58 percent) are producing wells of which 847 are gas*
> *wells.*

Dr. Snead reported that important technological advances have facilitated the ultra-deep drilling of natural gas wells. The average time to reach a depth of 17,000 feet for two East Texas deep wells drilled in the same structure decreased from 170 days to seventy days in the seventeen years between 1985 and 2002. Moreover, advances in computer technology have produced breakthroughs in reservoir modeling that "enable better estimates of the size and location of recoverable deposits."

Realizing the potential for the deep-well drilling of natural gas, the US Department of Energy's Office of Fossil Energy established a "Deep Trek" program to lower the cost and improve the efficiency of drilling commercially productive deep wells.[66] "Deep Trek" maintains its "Office of Fossil Energy" bias despite describing deep-well natural gas drilling as needing to penetrate rock structures that sound more like bedrock than sedimentary layers. The common wisdom remains that natural gas, like oil, is a "fossil fuel." For those who have any doubt that the "fossil fuel" theory

---

[66]   US Department of Energy, "'Deep Trek' and Other Drilling R&D," loc.cit, supra at note #62.

is the politically correct version of the origin of natural gas, visit the Energy Information Administration's "Energy for Kids" page, which explains how millions of years ago the remains of plants and animals decayed into organic material that became trapped in rocks until pressure and heat changed some of this organic material into coal, oil, and natural gas.[67]

The "Deep Trek" site points out a very steep curve involved in implementing the technology needed to produce the estimated 125 trillion cubic feet of natural gas resources estimated to lie beneath the continental United States at depths of 15,000 feet or deeper:

> *Tapping into this resource will be both technologically, daunting and expensive. For wells deeper than 15,000 feet, as much as 50 percent of drilling costs can be spent in penetrating the last 10 percent of a well's depth. The rock is typically hot, hard, abrasive, and under extreme pressure. Often, in deeper wells, it is not uncommon for the drill bit to slow to only two to four feet per hour at operating costs of tens of thousands of dollars a day and for a land rig and millions of dollars a day for deep offshore formations. And it is exceedingly difficult to control the precise trajectory*

---

[67] "Energy Kids," Administration, US Department of Energy, "Natural Gas Basics: How Was Natural Gas Formed?" at http://www.eia.gov/kids/energy.cfm?page=natural_gas_home-basics#naturalgasformation.

of a well when the drill bit is nearly three miles below the surface.

The "Deep Trek" project is currently financing advances in technology, including the development of the polycrystalline diamond drill bit, currently the industry standard for drilling into difficult formations. Scientists at the Energy Department's Sandia National Laboratories developed a "diffusion bonding" approach that allow drill bit manufacturers to adhere industrial-grade diamonds to the bit.

Deep-earth natural gas, like deep-water oil production, strongly supports the theory that the origin of oil is abiotic, not organic in nature. Global reserve estimates for natural gas have also increased, as geo-scientists realize the abundance of deep-earth natural gas – a development that once again challenges peak production assumptions. The National Oceanic and Atmospheric Administration of the US Department of Commerce estimates that oceans cover 71 percent of the earth's surface. Estimates of deep-earth natural gas global reserves should increase dramatically, as have estimates of deep-water oil reserves, as technological advances permit natural gas producers to explore economically at greater depths below the water surface and at greater distances out from the continental shelves. Truthfully, with 71 percent of the earth's surface largely unexplored, geo-scientists have no way to reliably estimate the quantity of deep-earth and deep-water hydrocarbon fuels the earth truly may hold.

DOCUMENTS AND PHOTOGRAPHS

Three separate record collections from the World War II era were examined to produce these exhibits:

1.   The Combined Intelligence Objectives Subcommittee
2.   Operation Paperclip Declassified Files
3.   U.S. Strategic Bombing Survey

Most of these exhibits are published here for the first time.

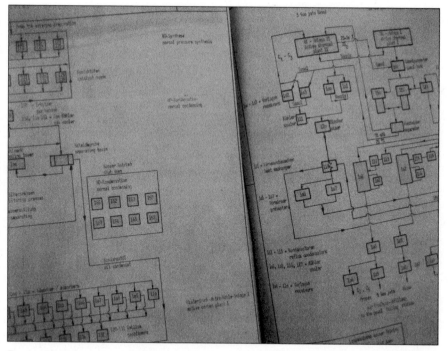

Declassified documents: schematic diagrams, Nazi FT plant construction, WWII combined intelligence assessments, Nazi FT plants. *Source: National Archives and Records Administration, Washington, D.C.*

4.0 SYNTHETIC OIL PRODUCTION

4.1 GENERAL

The outstanding feature of German oil economy during the past ten years has been the spectacular development of her synthetic oil plants for the production of oil from coal. This attempt at complete oil autarchy, made without regard to cost or orthodox financial considerations, has no parallel elsewhere and is a striking example of the character of the German master plan for world domination which called for the production, within her own boundaries, of all the resources essential to modern warfare. It is evident that one of the essentials in such a plan is the securing of adequate oil supplies and since the attempts to find natural petroleum deposits within her own borders met with a very limited success Germany naturally turned to other expedients. The complicated structure of the enormous synthetic oil industry has been built up, therefore, on the basis of political and strategic expediency, and on the foundation of Germany's wealth of coal deposits, especially of lignite or brown coal, as compared with her poverty in natural oil resources.

The extent to which the programme of synthetic production has been carried forward may be illustrated by the fact that approximately five out of every six gallons of gasoline and gas oil produced in Germany are derived not from oil wells, but from synthetic oil plants, and that the German synthetic production amounts to something like 60 per cent of total European (a) natural crude oil production.

A detailed history and an accurate economic appraisal of the synthetic oil industry is rendered difficult by the fact that, almost from its inception, the Germans realized the potential strategic importance of this industry, with the result that all but its broad outlines were closely shrouded in a cloak of secrecy, as were many features of their armament industries and other important elements of their national planning. Also, despite the rapid basic progress made in the prewar years, the greatest expansion in the synthetic industry actually has taken place since 1938. However, as a result of certain early commercial contracts a considerable amount of technical data were acquired from the Germans prior to the war which, supplemented by Allied aerial reconnaissance over the German synthetic plants themselves, has made possible fairly accurate appraisals of their processing methods and capacities.

What the synthetic program has cost the German nation, either in terms of monetary investment or of materials and manpower required for the construction and operation of the plants and the production of the required coal, has never been revealed. The structure of the industry is so complicated by government participation that it is difficult to estimate with any accuracy the capital investment in the synthetic oil industry or the cost of the synthetic oil produced. Both, however, are known to be enormous as compared to the cost of plant and production in the natural petroleum products industry. It has been estimated that the present German synthetic plants (b), having a total capacity of close to 5,000,000 metric tons of product per year, cost something like 4 or 5 billion Reichsmark or 1.6 to 2 billions of dollars. This is said to be from ten to thirty times the plant cost to produce similar quantities of liquid fuels from petroleum, depending upon the processes used.

By way of further comparison, prior to the war, the cost of a gallon of gasoline ex American refineries, excluding profits and taxes, was generally considered to be approximately 4 U.S. cents per gallon (adding some 2 cents for profits and shipping cost this gasoline could be layed down in Germany for about 6 cents per gallon), while the cost to manufacture a gallon of gasoline from coal by either of the major synthetic processes is at least 20 cents (c), or five times as great.

(a) Excluding Russia.
(b) The bare plant cost exclusive of mines, coke ovens, coal carbonization plants, or other ancillary or auxiliary processes.
(c) Approximately 200 Reichsmarks per ton.

Excerpt from the United States' Petroleum Facilities of Germany report, prepared by The Enemy Oil Committee for the Fuels and Lubricants Division Office of the Quartermaster General, regarding synthetic oil production, page 143. *Source: National Archives and Records Administration, Washington, D.C.*

In consideration of the foregoing, as well as for other reasons, the participation of the German petroleum companies, and particularly those with international affiliations, in the synthetic oil industry has been small. Rather, it is the
German coal, chemical, and heavy industries, under government direction and subsidy, which have been responsible for the development of synthetic plants and production.

From its earliest days the synthetic oil industry has been the subject of government encouragement and subsidies, and eventually and inevitably due to the magnitude of the program and the nature of the German state, to government direction
and control. All the experimental work with the process discovered by Professor
Bergius was carried on under the sponsorship of I.G. Farbenindustrie, and the second of
the two main synthetic processes was worked out by Professor Fischer and Dr. Tropsch
under the auspices of the Ruhr Coal Owners Association, but because of the heavy
investments required, industry was slow to embark on large scale commercial production. However, the leaders of the German coal, chemical, and heavy industries no
doubt realized the vital role these processes might play in any future war and pro-
ceeded with their development fully confident that any German government would,
sooner or later, foster their growth.

The advent of the Nazi government merely accelerated the development of this
and other German key industries by greatly increasing the already existing govern-
mental subsidies and direction. This trend came into full maturity with the inaug-
uration of the Four Year Plan under which all resources and industries were incor-
porated in a gigantic and strictly controlled production program subordinated to
national strategy, regardless of the usual commercial and economic considerations.
To carry out the ambitious and vital synthetic program, companies, in which the
coal, chemical, and heavy industries participated, were formed under State direction.
The State assisted by granting extensive and generous credits and subsidies, which,
in many cases, covered half the cost of new plant construction which from then on
was pushed with intensity. As pointed out under "Government Corporations " on page
13, all the companies in the industry must belong to the "trade association",
Wirtschaftsgruppe Kraftstoffindustrie, through which channel government instructions
to the industry are passed.

4.2 PRINCIPAL COMPANIES

Although German corporate structures are complex, the more important companies
that have been identified as engaged in the production of synthetic oil in Germany
are listed below. Further details on these and other companies may be found in the
German year book "Handbuch der Internationalen Petroleum-Industrie".

Braunkohle-Benzin A.G. (Brabag).- This company with head office at Berlin W8,
Schinkelplats 1/2, was formed in 1935, under State direction which required joint
participation by the various German brown coal (lignite) interests. The capital
stock is subscribed jointly by:

    A.G. Sächsische Werke, Dresden
    Anhaltische Kohlenwerke, Halle
    Braunkohlen- und Brikettindustrie A.G., Berlin
    Deutsche Erdöl A.G., Berlin
    Elektrowerke A.G., Berlin
    I. G. Farbenindustrie A.G.
    Ilse Bergbau A.G., Grube Ilse
    Mitteldeutsche Stahlwerke, Riesa
    Rheinische A.G. für Braunkohlen Bergbau und Brikettfabrikation, Cologne (Köln)
    Werschen-Weissenfelser Braunkohlen A. G., Halle

In June 1939, the capital of the company was RM. 100 million, and provision was
made for increasing this by RM. 25 million over the next five years. The value of
the plants already erected or under construction in 1938 was, according to the bal-
ance sheet, RM. 295 million.

Excerpt from the United States' Petroleum Facilities of Germany report,
prepared by The Enemy Oil Committee for the Fuels and Lubricants
Division Office of the Quartermaster General, regarding synthetic oil
production, page 144. *Source: National Archives and Records Administration,
Washington, D.C.*

Geheim!

An die
I.G.Farbenindustrie A.G.
z.Hd.v.Herrn Dir.Dr.Müller-Cunradi.
L u d w i g s h a f e n a.Rh.

Betr.: Errichtung der Flora-Anlage in Heydebreck.

Unter der Voraussetzung, dass die Gesamtinvestitionen
für die Errichtung der Flora-Anlage in Heydebreck in
Höhe von 24 Mio.RM von meinem Planungsamt genehmigt
werden, erkläre ich mich bereit, Ihnen die anfallende
Erzeugung von 400 mcto Bleitetraethyl für die Dauer von
5 Jahren nach Inbetriebnahme der Anlage nach Maßgabe
eines noch abzuschliessenden Liefervertrages abzunehmen.

Sollte eine vorzeitige Stillegung der Anlage in Erwägung
gezogen werden, so besteht Einigkeit darüber, dass Sie
keinerlei Anspruch auf entgangenen Gewinn bei der Ein-
stellung der Produktion stellen werden.

Ich weise darauf hin, dass über die Erzeugung der
Flora-Anlage nur mit meinem Einvernehmen (GL/AM) ver-
fügt werden darf. Die Abrechnungsunterlagen für das Bau-
vorhaben sind meiner Abteilung GL/AM zu gegebener Zeit
vorzulegen.

Ich bin bereit, Ihnen nach den geltenden Richtlinien
in den Preisen der Erzeugung folgende Abschreibungssätze
bei dreischichtigem Betrieb zu gewähren:

20 % auf Maschinen und Einrichtungen
7 % auf Gebäude und bauliche Anlagen (Umkleidungen

Ich habe zur Kenntnis genommen, dass Sie sich an der
Finanzierung des Bauvorhabens mit eigenen Mitteln in Höhe

- 2 -

Nazi air force ministry letter regarding IG Farben synthetic fuels war efforts, dated August 9, 1943, page 1. *Source: National Archives and Records Administration, Washington, D.C.*

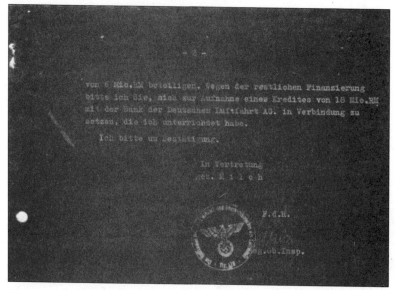

Nazi air force ministry letter regarding IG Farben synthetic fuels war efforts, dated August 9, 1943, page 2. *Source: National Archives and Records Administration, Washington, D.C.*

Nazi financing commitment for Fischer-Tropsch synthetic fuel. *Source: National Archives and Records Administration, Washington, D.C.*

OFFICE OF THE UNITED STATES HIGH COMMISSIONER FOR GERMANY
OFFICE OF ECONOMIC AFFAIRS
INDUSTRY DIVISION

30 December 1949

Mr. C. H. Nordstrom
Chief, Scientific Research Division
Military Security Board
APO 742, U.S. Army

Hermann Reusch
Fischer-Tropsch

Dear Mr. Nordstrom:

Attached is a cable received from Acheson, Secretary of State, together with a copy of the dispatch 473 to which the Secstate cable refers.

We also attach a copy we have sent forward to the Secretary of State indicating our lack of information on this subject. Could you take the necessary steps to investigate and make further reply.

Very truly yours,

H. A. TAYLOR
A/Chief
Industry Division

Encl:
1. Cable A-793
2. Bremen 473
3. Secstate Cable

Correspondence from the Office of the U.S. High Commissioner for Germany, dated December 30, 1949, documenting Secretary of State Dean Acheson's interest in Nazi FT plants. *Source: National Archives and Records Administration, Washington, D.C.*

There are ten factories in the British Zone for the production of synthetic oils. Six of them work according to the Fischer Tropsch process at atmospheric or low pressure, up to 10 atmospheres, and four of them work according to the Bergius or high pressure process, at from 300 - 700 atmospheres. All of them have been damaged to some extent as is described below.

Fischer Tropsch Process

1. **Krupp Treibstoffwerke, Essen.** This factory was damaged but has now been repaired and is capable of working at 60% capacity. It has been activated for the production of synthetic fats and a detergent, Mersol, both of which are urgently needed for the soap programme. Present capacity 38,000 tons/annum of hydrocarbons.

2. **Gew. Victor, Caustrop Rauxel.** This factory is in a state similar to Krupps, namely working at about 60% of the original capacity for the production of fatty acids and Mersol for the soap programme. Present capacity 24,000 tons/annum hydrocarbons.

3. **Essener Steinkohle.** This works is almost ready for production, which will be practically at its original capacity of 84,000 tons/annum hydrocarbons.

4. **Ruhr Chemie A.G.** War damage and cannibalisation have left capacity equal to about 50% of the original, that is to say, 39,000 tons/annum hydrocarbons.

5. **Dortmunder Paraffinwerke, Dortmund.** This factory is about 60% damaged and has a residual capacity of 25,000 tons/annum hydrocarbons.

6. **Steinkohlenbergwerk, Rheinpreussen.** This factory is about 30% damaged and has a residual capacity of 51,000 tons/annum hydrocarbons.

On normal running the Fischer Tropsch process produces the following:-

	Tons per 100 tons mixture	Value per ton £	Total Value £
Petrol	44	27.25	1199
Propane & Butane	11	27.25	300
Diesel Oil & Cosamin	32	19.75	632
Waxes	13	122.5	1595
		Total	3726

Value for one ton of mixed products £ 37.26

The above plants work on hard coal, 6 tons of which are necessary for the production of 1 ton of primary mixture. It is thus apparent that approximately 66 dollars worth of coal must be expended for 37.25 dollars worth of mixed products. The value given to the mixed products are the present import values and coal is costed at an export value of £11 per ton.

These factories can only be regarded as part of Germany's effort towards self-sufficiency and are uneconomical in normal conditions.

Inventory report on factories in the British zone capable of producing synthetic oils. *Source: National Archives and Records Administration, Washington, D.C.*

Nazi FT plants: diagrams, operational procedures, official documents.
Source: National Archives and Records Administration, Washington, D.C.

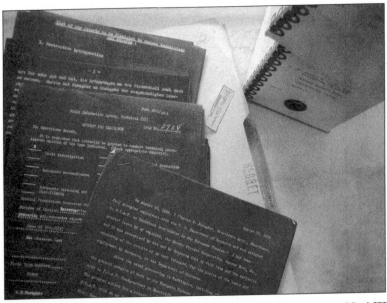

Declassified documents, WWII combined intelligence assessments, Nazi FT
plants. *Source: National Archives and Records Administration, Washington, D.C.*

Nazi FT scientist Helmut Pichler, fingerprint ID card, Operation
Paperclip declassified document. *Source: National Archives and Records
Administration, Washington, D.C.*

Joint Intelligence
Objectives Agency

10 January 1949

JIOA 104

MEMORANDUM FOR Mr. H. J. L'Heureux, Chief, Visa Division, Department of
State

SUBJECT: Case of Helmut PICHLER.

    1. Reference is made to JIOA letter 2375 dated 8 July 1948, addressed
to you.

    2. The subject alien, Helmut PICHLER, is hereby recommended by the
Joint Chiefs of Staff, as provided by Section 61.213 (a)(3)(1)(b) of Title
22, Code of Federal Regulations, as a person whose admission is highly
desirable in the national interest and is in accordance with the policy
pertaining to the entry of such aliens as established by the Joint Chiefs
of Staff.

ROBERT B. WEW
Captain, USN
Director

FILE DIST:
383.7 State Immigration
383.7 PICHLER, Helmut

104

Memorandum to H. J. L'Heureux, chief of Visa Divison, Department
of State, dated January 10, 1949, letter of recommendation for Helmut
Pichler. *Source: National Archives and Records Administration, Washington, D.C.*

Reference 4.

PATENTS AND PATENT-APPLICATIONS

DR. H. PICHLER

A. Synthesis of Benzene and Acetylene from Methane

　　1. Process for the production of higher hydrocarbons.
　　　DRP 643386.

　　2. Process for the production of hydrogen-poor unsaturated hydrocarbons from hydrogen-rich hydrocarbons.
　　　DRP 649102

　　3. Process for the production of carbon-rich hydrocarbons from carbon-poor hydrocarbons (partial combustion of hydrocarbons with formation of acetylene)
　　　DRP 553178

B. Normal Pressure Synthesis of Hydrocarbons from Carbon Monoxide and Hydrogen.

　　4. Improvement of yields of the catalytic synthesis of aliphatic hydrocarbons (Performance of the process in several steps). (Austria Patent 160916).

C. Medium Pressure Synthesis with Cobalt-Catalyst

　　5. Process for the production of paraffin from carbon monoxide and hydrogen.
　　　DRP 731295

　　6. Process for the production of hydrocarbons with the medium pressure synthesis on cobalt catalyst.
　　　Germany ST 56459

D. Medium Pressure Synthesis with Iron-Catalyst

　　7. Synthesis of higher hydrocarbons. (Germany ST 56470)

　　8. Process for the production of hydrocarbons from carbon monoxide and hydrogen. (Germany ST 56856)

　　9. Process for the production of higher hydrocarbons from carbon monoxide and hydrogen containing gases at higher pressures. DRP 738091

　　10. Process for the production of hydrocarbons from carbon monoxide and hydrogen (Germany ST 60409)

　　11. Process for the manufacture of iron-catalysts. (Germany ST 60795).

List of patents and patent-applications made by Helmut Pichler, page 1.
Source: National Archives and Records Administration, Washington, D.C.

E. Synthesis of High Molecular Weight and Previously Unknown Paraffins (Ruthenium-Catalysts)

    12. Process for the synthesis of solid aliphatic hydrocarbons. DRP 705528

F. Converter for the Synthesis of Hydrocarbons

    13. DRP 708500

G. Synthesis of Hydrocarbons in Liquid Phase

    14. DRP 716853

H. Synthesis of Formic-Acid from Carbon Monoxide and Water

    15. DRP ? (ST 61469)

I. Iso-Synthesis

    16. Process for the catalytic synthesis of branched hydrocarbons from carbon-monoxide and hydrogen. (ST 61125)

    17. Process for the catalytic synthesis of branched hydrocarbons from carbon-monoxide and hydrogen. (ST 62438)

    18. Process for the production of hydrocarbons with high octane-numbers. (ST 62589)

    19. Formation of branched hydrocarbons from dimethylether. (ST 62439)

    Most of the patents mentioned above have been granted or are claimed in many countries of the world.

List of patents and patent-applications made by Helmut Pichler, page 2.
Source: National Archives and Records Administration, Washington, D.C.

As a native born citizen of Austria I had no affiliation with any political group or party there. I went to Germany in 1927 to do post graduate research work. I was invited to do scientific work at Kaiser Wilhelm Institute of Coal Research at Muelheim-Ruhr by Dr. Franz Fischer, Director of that Institute. After I returned to Austria I received my Ph.D. from the University of Vienna in 1929. Dr. Franz Fischer then urged me to return to the Institute for Coal Research as his assistant. I did so becoming later Head of the Division for Synthetic Fuels; later I was made Deputy Director of the Institute.

Dr. Fischer, in 1932, urged me to become a citizen of Germany as a matter of policy of the Institute. Being a non-citizen of Germany prevented my technical advancement I also desired to lecture at Muenster University. After Dr. Fischer again asked me to become a German citizen I undertook to do so. My continuing research progress depended upon acquiring German citizenship. But my application to get citizenship depended on membership in the party so I became a member of it, the NS-Deutsche Arbeiterpartei, in 1933. The chief of our Institue was also a member. As scientists we were not concerned with politics.

In 1934, I became a citizen of Germany. Because I refused to participate in political drills I could not lecture at he Muenster University. When approached by the police, the SA and industry, through our Institute for Coal Research, with a demand to give lectures on air-raid protection, I could not refuse and held approx. ten lectures concerning civilian air defense (fighting incendiary bombs, etc.). For this I was given the title of SA-Truppfuehrer (squad leader), without ever having participated in other SA activities and without ever being the leader of a single man.

In 1943 I was requested to become active in the SA. I refused to do this. Thereupon I was threatened with expulsion from the SA. I did nothing further and was never approached thereafter. As were all other employees of the Institute for Coal Research and generally all German employees, I was a member of the Deutsche Arbeitsfront. I was a member of the NS Volkswohlfahrt, the oest. Hilfsbund, the Bund f. Leibesuebungen (automatically as a member of the Alpenverein and the Alpine Ski-und Tourenklub). I was not a member of these for political reasons.

All my thoughts and all my sympathies were ever concerned with my scientific work only. I performed this work in the same way before 1933, after 1933 and after 1945 See article in "The Saturday Evening Post" (October 6, 1945) where, with my picture, it is reported:

"German passion for scientific work regardless of temporary world conditions was further emphasized when the Kaiser Wilhelm Institute in Muelheim, Ruhr was reached. This government-financed research laboratory was the largest in Germany and had been responsible for some of the most valuable synthetic-fuel discoveries. Dr. Pichler, the acting director, was calmly supervising laboratory activities as if conditions were perfectly normal."

My desire to preserve the reputation of our institute for science was rewarded. In spite of Hitler and in spite of the war we continued our work. After arrival of American and British armies we obtained permission to continue our work in the same way as before.

I, the undersigned applicant, hereby swear (or affirm) that the above statements and data are true and correct to the best of my knowledge and belief.

_____
Signature of Applicant

Sworn to (or affirmed) and subscribed before me this 21st day of November 1947.

Notary Public, State of New York
Residing in Queens County
Queens County Clerk's No. 1216
Certificate filed in
N.Y. Co. CR No 424, Reg. No. 633-K-8
Commission expires March 30

_____
Title

See letters from Prof. Fischer and Prof. Hahn attached to list of professional references.

Nazi FT scientist Leonard Alberts, fingerprint ID card, Operation Paperclip declassified document. *Source: National Archives and Records Administration, Washington, D.C.*

R. D. 2
Library, Pa.
July 30, 1948

AFFIDAVIT

I, Leonard W. Alberts, presently under contract to Bechtel Corporation
as a "paperclip specialist", by reason of transfer of contract from the
War Dept., Office of the Quartermaster General, Washington , D.C. hereby
do affirm and swear under under oath the following facts:

    1. It might wellbe that I was listed as a registered member of the
       organisation , known as Foerderndes Mitglied (furthering furthering
       member) which has been stated to be a supporting league of the S. S.

       I have contributed to many National organisations of which this
       may have been one; however I do not have any specific recollection
       of membership in the aforementioned organisation because such
       matters were entrusted to my secretary ,Fraulein Edith Marzotko,
       presently employed by Ruhrchemie A.G. Oberhausen Holten, Germany
       (British Zone) who resides at Sterkrache-Buschhausen, X
       Pestalozzi St. 6. The aforementioned person was my personal
       secretary for the sixteen year period from 1930 through 1946.

    2. At the time of my first interrogation and to the best of my
       knowledge and belief, I stated that my statements may be partly
       incorrect due to loss of all my papers by an air raid in which my
       secretary's office was destroyed.

    3. In view of the foregoing statements contained in paragraphs 1
       and 2 above, I believe that sufficient information has been
       given regarding the possible ommission of mention of my registration
       as a member of the Foerderndes Mitglied. It was not intentional
       that such fact or facts, if true, be concealed from the interested
agenciesof the United States Government and the War Department.

SWORN AND SUBSCRIBED TO
UNDER OATH THIS
30TH DAY OF JULY 1948

Witnesses:

_____
H.T. Mc Bride

_____
R.A. Alberts

_____
R. E. Deckman

_____
Leonard W. Alberts

Sworn and subscribed before me this
30th day of July, 1948 at Library, Pa.

_____
Capt. C.M.F. Hartzel
Summary Court Officier

Nazi FT scientist Leonard Alberts, political affiliations affidavit,
Operation Paperclip declassified document. *Source: National Archives and
Records Administration, Washington, D.C.*

As Director of the Ruhrchemie A.G. in 1933 I was naturally pressed to affiliate myself with the N.S.D.A.P. It was possible for me in contrast to the other Directors of my firm to keep aloof from this membership.

In 1935 I was offered the position on the Board of Directors of the Briunkohle-Benzin A.G. However, after it had been determined that I was not a member of the N.S.D.A.P., this offer was withdrawn. In 1938 I got a similar offer from Krupp. This offer was also withdrawn for the same reason.

These two examples convinced me that without party membership I would not be able to accept offers which would improve my professional position. Therefore, I applied for a membership in 1938.

After becoming a member of the party in 1941, there were no further obstacles for me; so I could join the Board of Directors of the Gewerkschaft Victor in 1943.

My further applications with NS organizations consisted of: DAF, NSV, NSBDT in which I was only a paying member.

In 1934 I joined an engineering division of the SA in order to relieve myself from mental strain through physical activities. This division was dissolved after a short time and I was transferred to a non-engineering division of the SA. Since I was not interested in the regular SA, I continuously refused to follow their requests for service, until I was finally expelled in 1935.

I, the undersigned applicant, hereby swear (or affirm) that the above statements and data are true and correct to the best of my knowledge and belief.

_____
(Signature of Applicant)

SWORN TO (OR AFFIRMED) AND SUBSCRIBED BEFORE ME THIS

__21__ day of ___Oct.___ 1947.

_____
(Title)

(SEAL)

MARTHA LAFLANDER MULLEHAUSEN
MY COMMISSION EXPIRES
FEBRUARY 21, 1949

Nazi FT scientist Leonard Alberts, political statement, Operation Paperclip declassified document. *Source: National Archives and Records Administration, Washington, D.C.*

Department of Justice
Office of the Assistant to the Attorney General
Washington
November 9, 1949.

Colonel Daniel E. Ellis, USAF   CONFIDENTIAL
Director, Joint Intelligence Objectives Agency
Room 2D-880, The Pentagon
Washington 25, D. C.

Dear Colonel Ellis:

    Reference is made to the several communications
between your office and this Department regarding Leonard Wilhelm
Alberts, a German scientist now in this country under the Paperclip Program.

    You have previously been furnished with pertinent
portions of Federal Bureau of Investigation reports concerning Alberts.
In view of derogatory information contained therein conferences have
been held between members of your staff and representatives of this
Department and subsequent communications have been furnished from the
Quartermasters Corps, United States Army, and the Blaw-Knox Company
regarding Alberts.

    Upon consideration of all the information received
concerning Alberts this Department is of the opinion that it cannot
recommend him to the Immigration and Naturalization Service for permanent admission into the United States. You will note that Alberts
served for a time during World War II as a functionary of the Abwehr,
the German Intelligence. The statements of several persons who have
known Alberts, including Major Robert E. Humphries, who has been
directly concerned with security matters pertaining to the presence
of German scientists at Bureau of Mines plants, have grave misgivings
of Alberts as a security risk. It would appear that he is pro-Nazi
in his outlook and unscrupulous in his activities and, as Major Humphries has stated, he is capable of dealing with Russia or any other
group which would pay for his technical knowledge. In this connection
it may be pointed out that the Blaw-Knox Company for which he now works
plans to send him to Germany, a factor which has a definite bearing on
the security risk in admitting Alberts.

    In view of the foregoing, the Department of Justice is
of the opinion that it cannot agree that Alberts' presence in this country would not constitute a risk to the internal security. If you have

4181

Nazi FT scientist Leonard Alberts, U.S. Attorney General declines
support, letter dated November 9, 1949, page 1, Operation Paperclip
declassified document. *Source: National Archives and Records Administration,
Washington, D.C.*

any other information which you wish to call to the Department's
attention in this matter, please do not hesitate to communicate
with me.

Yours sincerely,

Peyton Ford
The Assistant to the Attorney General

FILE DIST:
ALBERTS, Leonard Wilhelm
C/R Justice Immigration

4161

Nazi FT scientist Leonard Alberts, U.S. Attorney General declines
support, letter dated November 9, 1949, page 2, Operation Paperclip
declassified document. *Source: National Archives and Records Administration,
Washington, D.C.*

THE SECRETARY OF COMMERCE
WASHINGTON 25

The Honorable
The Attorney General
Department of Justice
Washington 25, D. C.

JUL 14 1950

Dear Mr. Attorney General:

This will supplement my letter of September 28, 1948 concerning Leonard W. Alberts, a German specialist in the field of synthetic fuels.

In the earlier letter I advised that Alberts' technical background, accomplishments and professional standing indicate that his presence in this country would be in the interest of our national economy. We have recently been informed by his employers, the Blaw-Knox Construction Company, that their efforts to date to obtain Alberts' immigration visa have been unsuccessful. In their letter of June 30, 1950 to me, copy attached, they state that Alberts has been employed by them for eighteen months and that during this period they have been favorably impressed by his abilities and by his personal attitude. This belief is reinforced by supplemental correspondence from Mr. C. R. Breck of the Southern Natural Gas Company and Mr. Donald W. Deery, senior chemical engineer of Blaw-Knox.

The Fischer-Tropsch process for the production of synthetic fuels, in which Alberts is expert, may be a significant item in our national defense. In this connection I am also bringing to your attention a copy of a letter recently received from Mr. Storch of the Bureau of Mines. As you will also note, personnel in the central experimental station of the Bureau of Mines regularly discuss synthetic fuel developments and related technical matters with Mr. Alberts.

Since Alberts recently went to Germany under Army supervision, it was necessary to provide a reliable custodian. Mr. Donald Deery of Blaw-Knox accomplished this function. Therefore his letter giving certain facts concerning Mr. Alberts seems most pertinent.

Normally it would appear that the significance of the Fischer-Tropsch development to our economy would be adequate to request a careful consideration of Mr. Alberts' application for immigration, based on facts furnished herewith. The critical national situation would seem to reinforce this request. Synthetic fuels are vital to military operations, also Mr. Alberts' fund of information would seem to make it more desirable for him to stay in this country than to return to Germany.

Sincerely yours,

THOMAS C. BLAISDELL, JR.

Acting Secretary of Commerce

Nazi FT scientist Leonard Alberts, U.S. Secretary of Commerce endorses as critical to "national defense," letter dated July 14, 1950, Operation Paperclip declassified document. *Source: National Archives and Records Administration, Washington, D.C.*

Nazi FT Scientist Erich Frese, fingerprint ID card, Operation Paperclip declassified document. *Source: National Archives and Records Administration, Washington, D.C.*

STATE OF MISSOURI)
) ss.
County of Pike   )

        Be it remembered that on this _twenth_ day of October in the
year of our Lord One Thousand Nine Hundred Forty Seven and before me,
the subscriber has personally appeared known to me as Erich Frese.

                    POLITICAL AFFILIATION

        I was a member of:

        1. "Deutsche Arbeitsfront" (D.A.F) from 1934 till March 1945.
        2. "Nationalsocialistische Volkswohlfahrt" (N.S.V.) from 1937
           till March 1945.
        3. "Nat. Soc.Reichsbund für Leibesübungen" from 1937 - 1938.
        4. N.S.DA.P. from April 1942 - March 1945.

        I was forced to join the N.S.D.A.P. in consequence of my de-
nomination as the "Acting Works Director" of the Ruhroel G.M.B.H. in
1942 only, because as a "Non-Party Member" I was liable continuously
to grave difficulties in personal regard as well as in my commitments
in charge of the management of the works.

        During my activity as works manager I was very often in opposition
to the party officials followed by grave arguments with them. Two days
before the Ruhroel G.M.B.H. was occupied by Allied Troops, I received by
special messenger from the Defense Commissioner, Ganleiter Meyer, the
strict order to demolish completely the factory, to burn the oil storage
(1800 t) and thereafter join the retreating German Army. I had to certify
with my signature that I am personally responsible for the fulfillment of
this order.

        I did not fulfill this order. I remained in the factory to be at
the disposition of the Allied Troops.

        I, _Erich Frese_, do swear (affirm) that I know the contents
of this affidavit and the same are true and correct.

                                        Erich Frese
                                            (Signed)

        Subscribed and sworn to before me this _20th_ day of _October_,
Anno Domini 1947.

                                        Charles W. Van Noy
                                            Notary Public
                                MY COMMISSION EXPIRES OCT. 8, 1950

Nazi FT Scientist Erich Frese, political affiliations statement, Operation
Paperclip declassified document. *Source: National Archives and Records
Administration, Washington, D.C.*

Nazi FT Scientist Hanns Schappert, fingerprint ID card, Operation Paperclip declassified document. *Source: National Archives and Records Administration, Washington, D.C.*

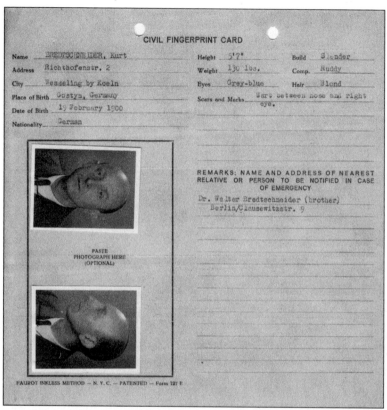

CIVIL FINGERPRINT CARD

Name BREDTSCHNEIDER, Kurt

Address Richthofenstr. 2

City Wesseling by Koeln

Place of Birth Gostyn, Germany

Date of Birth 19 February 1900

Nationality German

Height 5'7"

Weight 130 lbs.

Eyes Grey-blue

Build Slender

Comp. Ruddy

Hair Blond

Scars and Marks Wart between nose and right eye.

PASTE
PHOTOGRAPH HERE
(OPTIONAL)

REMARKS: NAME AND ADDRESS OF NEAREST
RELATIVE OR PERSON TO BE NOTIFIED IN CASE
OF EMERGENCY

Dr. Walter Bredtschneider (brother)
Berlin/Clausewitzstr. 9

FAUROT INKLESS METHOD — N. Y. C. — PATENTED — Form 127 E

Nazi FT Scientist Kurt Bredtschneider, fingerprint ID card, Operation Paperclip declassified document. *Source: National Archives and Records Administration, Washington, D.C.*

CIVIL FINGERPRINT CARD

Name Josenhans, Max   Height 5'0"   Build Medium

Address Stuttgart, Herresh Str. 27   Weight 120 lbs   Comp. Ruddy

City Blaubeuren   Eyes Blue   Hair Grey

Place of Birth Wildbad Kr. Calu   Scars and Marks Scar on hip

Date of Birth 24.12.1893

Nationality German

REMARKS: NAME AND ADDRESS OF NEAREST RELATIVE OR PERSON TO BE NOTIFIED IN CASE OF EMERGENCY

Klaus Josenhans
Leverkusen/Schlebusch 1
Stixchen str. 119

PASTE PHOTOGRAPH HERE (OPTIONAL)

FAUROT INKLESS METHOD — N.Y.C. — PATENTED — Form 127 B

25. List any German University Student Corps to which you have ever belonged. — 26. List (giving location and dates) any Napola, Adolph Hitler School, Nazi Leaders College or military academy in which you have ever been a teacher. — 27. Have your children ever attended any of such schools? Which ones, where and when? — 28. List (giving location and dates) any school in which you have ever been a Vertrauenslehrer (formerly Jugendwalter).

25. Welchen deutschen Universitäts-Studentenburschenschaften haben Sie je angehört?

26. In welchen Napola, Adolf-Hitler-, NS-Führerschulen oder Militärakademien waren Sie Lehrer? Anzugeben mit genauer Oris- und Zeitbestimmung

27. Haben Ihre Kinder eine der obengenannten Schulen besucht? Welche, wo und wann?

28. Führen Sie (mit Orts- und Zeitbestimmung) alle Schulen an, in welchen Sie je Vertrauenslehrer (vormalig Jugendwalter) waren.

C. PROFESSIONAL OR TRADE EXAMINATIONS / C. Berufs- oder Handwerksprüfungen

Name of Examination Name der Prüfung	Place Taken Ort	Result Resultat	Date Datum

Nazi FT Scientist Max Josenhaus, fingerprint ID card, Operation Paperclip declassified document. *Source: National Archives and Records Administration, Washington, D.C.*

General Introduction

Nature of the Target

The Ruhrchemie A.G. was formed by a group of Ruhr Companies, the principle shareholders being :-

Gelsenkirchener Bergwerke A.G.
Gutehoffnungshütte
Harpener Bergbau A.G.
Fried. Krupp A.G.
Mannesmann Röhren-Werke A.G.
Essener Steinkohlen Bergwerke A.G.
Rheinpreussen.

About 1935, following the acquisition by Ruhrchemie of exclusive rights to the Fischer-Tropsch process, the Company started the construction of a plant for the commercial operation of this process, and a subsidiary company known as Ruhrbenzin A.G. was formed to serve as the operations organisation for the project. It is believed that the desire to find outlets for the surplus coke produced in the Ruhr was one of the main reasons for the new venture.

Not only did this plant represent the initial large-scale operation of this process, but it also served as the research and development centre for the process and many modifications and ancilliary and related developments which followed later. The other eight Fischer-Tropsch plants which were subsequently erected in Germany were licensed by Ruhrchemie and were based on the Ruhrchemie model.

The catalyst factory at Sterkrade-Holten supplied all catalysts required for the operation of the six Fischer plants in the Ruhr area.

Plants erected (or projected) in foreign countries (e.g. Japan) were licensed by Ruhrchemie and depended on them for the basic technical information necessary for the operation of the process.

It is therefore clear that the Sterkrade-Holten works constituted a target of major importance for C.I.O.S. Item 30, and required thorough investigation. Other Fischer-Tropsch plants in Germany would serve mainly to confirm and amplify information obtained from Ruhrchemie.

Declassified Combined Intelligence Objectives Subcommittee (CIOS) report on Fischer-Tropsch plant of Ruhrchemie A.G. at Sterkrade-Holten. *Source: National Archives and Records Administration, Washington, D.C.*

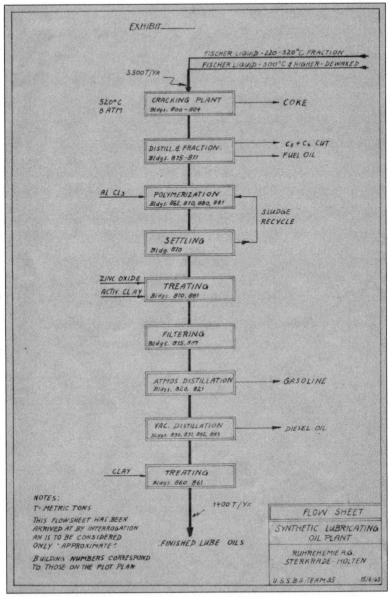

Declassified Combined Intelligence Objectives Subcommittee (CIOS) report on Fischer-Tropsch plant of Ruhrchemie A.G. at Sterkrade-Holten, flow chart of operations. *Source: National Archives and Records Administration, Washington, D.C.*

SYNTHETIC OIL PLANT STERKRADE/HOLTEN

Notes for Preparation of Bomb Data Information:

The plant area was 423.677 acres. At the direction of the technical team leader, physical damage of the Sterkrade/-Holten plant was confined to three specific areas: (1) Boiler House area 110 X 100 metres, (2) Catalyst Plant 34 X 90 metres, (3) Fischer-Tropsch Processing Tank Farm and incidental buildings in the south center portion of the plant covering areas 160 X 200 metres and 220 X 400 metres. In the areas designated above a total of 130 bombs were plotted, and the damage resulting from the explosion of these bombs was noted and analyzed.

Declassified Combined Intelligence Objectives Subcommittee (CIOS) report on Fischer-Tropsch plant of Ruhrchemie A.G. at Sterkrade-Holten, bomb damage assessment. *Source: National Archives and Records Administration, Washington, D.C.*

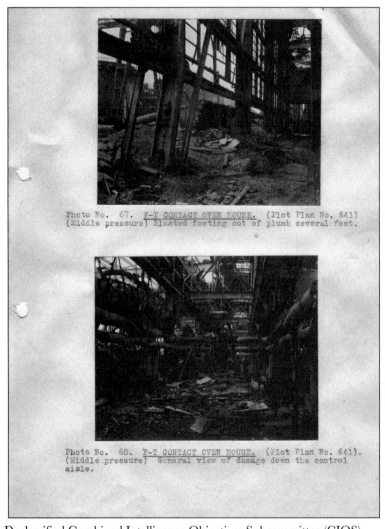

Photo No. 67. F-T CONTACT OVEN HOUSE. (Plot Plan No. 641)
(Middle pressure) Blasted footing out of plumb several feet.

Photo No. 68. F-T CONTACT OVEN HOUSE. (Plot Plan No. 641).
(Middle pressure) General view of damage down the control
aisle.

Declassified Combined Intelligence Objectives Subcommittee (CIOS) report on Fischer-Tropsch plant of Ruhrchemie A.G. at Sterkrade-Holten, bomb damage assessment. *Source: National Archives and Records Administration, Washington, D.C.*

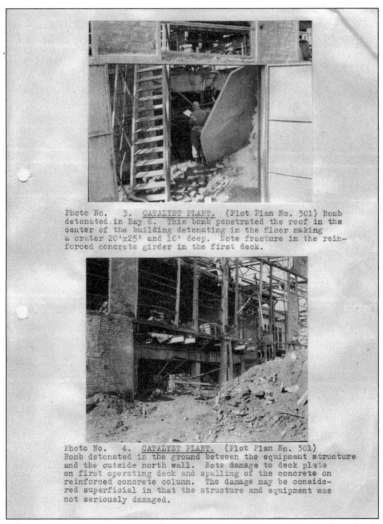

Photo No. 3. CATALYST PLANT. (Plot Plan No. 301) Bomb
detonated in Bay 6. This bomb penetrated the roof in the
center of the building detonating in the floor making
a crater 20'x25' and 10' deep. Note fracture in the reinforced concrete girder in the first deck.

Photo No. 4. CATALYST PLANT. (Plot Plan No. 301)
Bomb detonated in the ground between the equipment structure
and the outside north wall. Note damage to deck plate
on first operating deck and spalling of the concrete on
reinforced concrete column. The damage may be considered superficial in that the structure and equipment was
not seriously damaged.

Declassified Combined Intelligence Objectives Subcommittee (CIOS)
report on Fischer-Tropsch plant of Ruhrchemie A.G. at Sterkrade-
Holten, bomb damage assessment. *Source: National Archives and Records
Administration, Washington, D.C.*

Photo 62. Building 13A. This photo was taken from the ground floor and shows the converters that have been overturned by the blast effect from the 4,000-lb. bomb described in the preceding caption.

Photo 63. Building 13B. Another 4,000-lb. HE bomb detonated in the roof section of the Fischer-Tropsch reactor house shown above. Blast damage structurally destroyed the center of the building and the main piping loop as shown in the foreground.

CONFIDENTIAL

Declassified Combined Intelligence Objectives Subcommittee (CIOS) report on Allied bomb damage to Nazi Fischer-Tropsch plants. *Source: National Archives and Records Administration, Washington, D.C.*

Photo 56. A general view of the southern section of Lutzkendorf plant taken from the roof of boiler house 19. The gas purification units are shown at the left.

Photo 57. The Fischer-Tropsch catalyst manufacturing plant. This is a general view looking east at the catalyst plant known as Kartor-Fabrik-Lutzkendorf.

Declassified Combined Intelligence Objectives Subcommittee (CIOS) report on Allied bomb damage to Nazi Fischer-Tropsch plants. *Source: National Archives and Records Administration, Washington, D.C.*

MEMO TO: DIRECTOR, FOREIGN PROPERTY DIVISION

FROM: Director, Property Liquidation Division

SUBJECT: Status of the Fischer-Tropsch Patents

    1. According to available information, Japanese rights to the German Fischer-Tropsch hydro-carbon synthesis process for manufacturing synthetic fuels were acquired in 1936 by Mitsui from the German firm Ruhrchemie A.G.. The following Japanese plants were built during the war with the aid of Ruhrchemie technicians for the purpose of manufacturing by this process:

        a. Nippon Jinzo Sekiyu (old name)
           Amagasaki Shokuen Seizo K.K. (old name)
           Amagasaki, Hyogo Prefecture

        b. Takikawa Kagaku Kogyo K.K.
           Takikawa, Hokkaido

        c. Miike Gosei Kogyo K.K.
           Omuta, Fukuoka Prefecture

    2. It is requested that an investigation be initiated to determine whether there is any vested interest due in the way of royalties on patents, trade marks, and utility models; or license fees, payments for "know how", or any other funds from Mitsui or others to the account of Ruhrchemie A.G.

    3. It is not known by Enemy Property Branch which of the Mitsui interests originally acquired the patents.

                          F. A. MORRISON
                          Director, Property Liquidation Division

FILED

Declassified Combined Intelligence Objectives Subcommittee (CIOS) report on Nazi Fischer-Tropsch plants built in Japan. *Source: National Archives and Records Administration, Washington, D.C.*

5

"Julian Simon Says" – Toward a Comprehensive Energy Policy

In 1865, Englishman William Stanley Jevons, one of the greatest social scientists of his day, wrote an exhaustive study entitled *The Coal Question: An Inquiry Concerning the Progress of the Nation, and the Probable Exhaustion of our Coal Mines*.[68] Jevons argued that England was about to exhaust all available coal resources, collapsing the industrial enterprise

---

[68] William Stanley Jevons, *The Coal Question: An Inquiry Concerning the Progress of the Nation, and the Probable Exhaustion of our Coal Mines* (London: Macmillan and Company, 1866, Second Edition), Library of Economics and Liberty, at http://www.econlib.org/library/YPDBooks/Jevons/jvnCQ.html.

upon which Great Britain's mighty empire depended. He wrote:

> *It will appear that there is no reasonable prospect of any relief from a future want of the main agent of industry* [i.e., coal].[69]

And:

> *We cannot long continue our present rate of progress. The first check for our growing prosperity, however, must render our population excessive.*[70]

In contemplating his form of the Malthusian nightmare, W. Stanley Jevons was the "peak production" theorist of his day. His work is filled with detailed analyses of coal mines showing depletion rates, with mine-by-mine estimates of the amount of coal available, the annual production of that coal, and the maximum duration of the supply, anticipating with uncanny precision the "bell-shaped curve" typical of M. King Hubbert's "peak oil" graphs.

In his classic 1996 book, *The Ultimate Resource 2*, debunking a myriad of "doom-and-gloom" resource scares that abound in popular and scientific thinking, then-University of Maryland professor Julian L. Simon, explained why Jevons was flat wrong:

> *What happened? Because of the perceived future need for coal and because of the potential profit in meeting*

---

[69]   Ibid., "Introduction and Outline," http://www.econlib.org/library/YPDBooks/ Jevons/jvnCQ1.html#Chapter%201.

[70]   Ibid·

that need, prospectors searched out new deposits of coal, inventors discovered better ways to get coal out of the earth, and transportation engineers developed cheaper ways to move the coal.[71]

Insightfully, Julian Simon documented a series of authoritative predictions dating back to 1885, all warning that the US would soon run out of oil:

- 1885, US Geological Survey: "Little or no chance for oil in California."

- 1991, US Geological Survey: Same prophecy by USGS for Kansas and Texas as in 1895 for California.

- 1914, US Bureau of Mines: Total future production limit of 5.7 billion barrels of oil, at most a ten-year supply remaining.

- 1939, Department of the Interior: Oil reserves in the US to be exhausted in 13 years.

- 1951, Department of the Interior, Oil and Gas Division: Oil reserves in the US to be exhausted in 13 years.[72]

---

[71] Julian Simon, "When Will We Run Out of Oil? Never!" *The Ultimate Resource 2* (Princeton, NJ: Princeton University Press, 1996), pp. 162-181, at p. 165.

[72] Ibid.

When did Julian Simon think we would run out of oil? "Never!" was his answer. With 1.28 trillion barrels of oil in proven reserves today, more than ever in recorded human history (despite oil consumption in the world nearly doubling in the last three decades), we should seriously consider that Julian Simon might well be right.

"Peak Production" believers regard Shell Oil geologist M. King Hubbert as their theoretical deity. As noted earlier, in 1956, Hubbert drew a bell-shaped curve that he said showed US oil production would peak in the 1970s and decline from there until US oil would in 2050 be nearly depleted. Subsequently, Hubbert's adherents have expanded his analysis into a worldwide prediction that we are running out of oil. Again, "Hubbert's Peak" theorists have serious critics, including prominent oil and gas analyst Michael C. Lynch.[73] In a paper entitled "The New Pessimism about Petroleum Resources: Debunking the Hubbert Model (and Hubbert Modelers),"[74]Lynch argues that Hubbert's initial analysis was anything but rigorous or scientifically formal:

> *The initial theory behind what is now known as the Hubbert curve was very simplistic. Hubbert was simply trying to estimate approximate resource*

---

[73] Michael Lynch, President and Director of Global Petroleum Service, Strategic Energy & Economic Research Inc. (SEER), at http://www.energyseer.com/MikeLynch.html.

[74] http://www.energyseer.com/NewPessimism.pdf

levels, and for the lower-48 US he thought a bell-curve would be the most appropriate form. It was only later that the Hubbert curve came to be seen as explanatory in and of itself, that is, geology requires that production should follow such a curve. Indeed, for many years, Hubbert himself published no equations for deriving the curve, and it appears that he only used a rough estimation initially. In his 1956 paper, in fact, he noted that production often did not follow a bell curve. In later years, however, he seems to have accepted the curve as explanatory.[75]

Julian Simon argued that gloomy predictions about running out of oil, coal, or any other energy resource including natural gas, were typically wrong for several reasons:

- Typically all energy resources exist on earth in quantities much larger than initially estimated;
- Advances in technology make exploration and recovery of previously difficult to develop energy resources more efficient and economically affordable;
- Improvements in productivity lead to more efficient use of energy resources over time;
- Alternative sources of energy are found, even while predominately used energy resources remain abundant.

---

[75] Michael Lynch, "The New Pessimism about Petroleum Resources: Debunking the Hubbert Model (and Hubbert Modelers)," at http://www.energyseer.com/NewPessimism.pdf.

- Previously dominant energy resources, such as coal, become less dominant as more efficient energy resources, such as oil, become more understood and utilized – a process Simon believed would continue as liquefied natural gas replaces oil applications, culminating with nuclear energy replacing many current applications of oil and natural gas.

Simon's energy resource analysis maintains that we will be running automobiles with nuclear batteries long before we run out of oil. Nuclear power, according to Simon, is the final inexhaustible energy resource. Today, the US Navy runs ships around the world predominately on nuclear power, without any history of life- or environmental-threatening accidents. Simon wrote: "Of course nuclear power can replace coal and oil entirely, which constitutes an increase in efficiency so great that it is beyond my powers to portray the entire process on a single graph based on physical units."[76]

The example environmentalists and radical global warming alarmists typically neglect is France, a country that since the 1980s has built a network of modern nuclear power plants needed to power France's major cities for the foreseeable future. Today, approximately 80 percent of France's electricity is generated by fifty-nine nuclear plants across the country,

---

[76] Ibid., p. 177.

plants that are at least a generation more advanced than the nuclear power plants operating today in the United States.[77]

Simon concedes that just because we've proved previous predictions of oil depletion wrong does not prove that every gloomy forecast about oil will be wrong. He granted that forecasts could be overly optimistic, as well as overly pessimistic. But, he reminded us, history shows that expert forecasts about running out of hydrocarbon fuels have typically been far too pessimistic. After over a century-and-a-half of aggressively using coal worldwide, and nearly a century of aggressively using oil worldwide, we still have ample reserves and ready supplies of both globally. Simon cautioned, therefore, that we should be careful not to allow energy scarcity predictions to scare us.[78]

In answering the question "Why do we believe so much false bad news about the energy?" Simon explained that people have a tendency to see energy resources as finite. "The idea is found in Malthus, of course," he wrote. "But the idea probably has always been a staple of human thinking, because so much of our situation must sensibly be regarded as fixed in the short run – the number of bottles of beer in the refrigerator, the size of our paychecks, and the amount of energy that parents have to play basketball with their kids." In contrast, Simon felt it made more

---

[77] "Nuclear Energy: Power struggle. Will France continue to lead the global revival of nuclear power?" *The Economist*, December 4, 2008, at http://www.economist.com/business/displaystory.cfm?story_id=12724850.

[78] Simon, "Will We Run Out of Oil? Never!" op.cit., p. 165-166.

sense to see energy as a fixed resource, not a finite resource.[79] The history of hydrocarbon fuels confirms Simon's viewpoint. If Jevons had been right, we would have been out of coal long ago. If Hubbert had been right, there would be no need for gas stations because there would be virtually no gasoline left to fuel our vehicles anywhere in the world, regardless of how much we might be willing to pay per gallon to fill up our gas tanks.

Obama Bans Offshore Drilling in Favor of Offshore Wind Power

The Obama administration has openly displayed an ideological preference for green energy, despite abundant evidence that green energy technologies, including wind and solar, fail to deliver the robust energy supply the United States needs to sustain strong economic growth.

On Dec. 1, 2010, Interior Secretary Ken Salazar announced a seven-year moratorium on offshore oil exploration into the eastern Gulf of Mexico along the Atlantic Coast, as a result of the Deepwater Horizon disaster. "As a result of the Deepwater Horizon oil spill, we learned a number of lessons," the *New York Times* quoted Salazar as saying in a press briefing at the time, "most importantly that we need to proceed with caution and focus on creating a stringent

---

[79] Julian Simon, "The concepts That Lead to Scares about Resources and Population Growth," in Julian Simon, *Hoodwinking the Nation* (New Brunswick, USA., and London, U.K.: Transaction Publishers, 1999), pp. 31-44, at pp. 41-33.

regulatory regime."[80] After the BP spill, Salazar closed the Minerals Management Service, the regulatory agency whose laxness the Obama administration blamed for the oil rig explosion, and replaced it with a new regulatory agency charged with performing more regular and rigorous oil rig inspections and enforcement of environmental and safety rules. In a press release announcing the decision, Salazar explained that the moratorium was intended to provide time to get the new regulatory structure in place.[81]

Dr. Joseph Mason, the chair of banking at the Louisiana State University and a well known economist, has estimated that the offshore drilling moratorium imposed by the Obama administration immediately after the Deepwater Horizon oil spill would cost the Gulf Coast a loss of 8,000 jobs and $500 million in lost wages in the first six months. "The moratorium could be more costly than the oil spill itself," Mason told reporters.[82]

Only a few days before the moratorium went into effect, President Obama directed his Interior Department to facilitate leases for offshore wind turbines. On Nov. 23, 2010, Salazar

---

[80] John M. Broder and Clifford Krauss, "US Halts Plan to Drill in Eastern Gulf," *New York Times*, Dec. 1, 2012, at http://www.nytimes.com/2010/12/02/us/02drill.html?pagewanted=all.

[81] Press Release, "Salazar Announces Revised OCS Leasing Program," US Department of the Interior, Dec. 1, 2010, at http://www.doi.gov/news/pressreleases/Salazar-Announces-Revised-OCS-Leasing-Program.cfm.

[82] Rebecca Torrellas, "UPDATED: Offshore drilling moratorium affects everyone," EPmag.com, July 19, 2020, at http://www.epmag.com/Production-Drilling/UPDATED-Offshore-drilling-moratorium-affects-everyone_63533.

announced his department's intentions to simplify and speed up the process of applying for and obtaining offshore leases for wind energy. Applying a value-loaded title to the program, Salazar announced in a Department of Interior press release that the "Smart from the Start" wind energy initiative for the Atlantic Outer Continental Shelf was designed "to facilitate siting, leasing and construction of new projects, spurring the rapid and responsible development of this abundant resource."[83]

The "Smart from the Start" wind power initiative was a follow-up to the Cape Wind project Salazar had announced only two months earlier. On Oct. 6, 2010, Salazar signed the nation's first lease for commercial wind energy development with Cape Wind Associates, LLC, a subsidiary of Energy Management, Inc.[84] The area involved in the Cape Wind project comprised twenty-four square miles of the Outer Continental Shelf in the Nantucket Sound off the shores of Massachusetts. The 130 planned wind turbines each had a hub height of 285 feet. The footprint for the Cape Wind project site is about five miles from the mainland shore, thirteen miles from Nan-

---

[83] Press Release, "Salazar Launches 'Smart from the Start' Initiative to Speed Offshore Wind Energy Development off the Atlantic Coast," US Department of the Interior, Nov. 23, 2010, at http://www.doi.gov/news/pressreleases/Salazar-Launches-Smart-from-the-Start-Initiative-to-Speed-Offshore-Wind-Energy-Development-off-the-Atlantic-Coast.cfm.

[84] Press Release, "Salazar Signs First US Offshore Commercial Wind Energy Lease with Cape Wind Associates, LLC," US Department of the Interior, Oct. 6, 2010, at http://www.doi.gov/news/pressreleases/Salazar-Signs-First-US-Offshore-Commercial-Wind-Energy-Lease-with-Cape-Wind-Associates-LLC.cfm.

tucket and nine miles from Martha's Vineyard. At peak power, the offshore wind farm is estimated to generate a maximum electric output that could produce enough energy to power approximately 420,000 homes. The Interior Department further estimated that the Cape Wind energy project could generate enough power to meet 75 percent of the electricity demand for Cape Cod, Martha's Vineyard, and Nantucket Island combined.

"One-fifth of the offshore wind energy potential is located off the New England coast, and Nantucket Sound receives strong, steady Atlantic winds year round," the Interior Department press release announcing the Cape Wind project noted. The Interior Department suggested that the Bureau of Ocean Energy Management, Regulation and Enforcement was expected to begin issuing new offshore leases for wind turbine power by the end of 2011, under the streamlined process. When announcing the Cape Wind project, the Interior Department also made public that the agency was considering offshore wind energy leases along the Outer Continental Shelf of Maryland, Delaware, New Jersey, Virginia, and Rhode Island, in addition to Massachusetts.

Famously, the late Massachusetts Democratic Senator Ted Kennedy rigorously objected for years to putting wind turbines offshore on Cape Cod because he felt the damage done by the wind turbines to the scenery of Cape Cod outweighed the value of obtaining the wind-turbine green energy. Approximately a year and a half after the Interior Department announcement,

Cape Wind finally selected three private contractors to build the wind power facility off Nantucket Island, after fierce community debate from residents of Cape Cod, Martha's Vineyard, and Nantucket because of the expected impact on the scenic beauty of the offshore area as well as fears that the electricity generated at the wind farm would raise prices. Cape Wind further announced that construction of the offshore wind farm was not expected to begin until 2013.[85] Current estimates are that the Cape Wind project will be more expensive than generating electricity through hydrocarbon fuels, with the Cape Wind project expected to add $1.08 to the monthly bill of the average residential customer in the Bay State.[86]

Oilman T. Boone Pickens Wind Farm Plan Goes Bust

If anything indicates that wind turbine energy is not yet a large-scale commercial energy technology, it's decision by renowned oilman T. Boone Pickens to abandon his proposed billion dollar wind farm in Pampa, Texas, a small town in the Texas panhandle. If anyone could have been expected to make wind turbine energy work in a commercially viable operation, it was Pickens.

---

[85] "Cape Wind picks contractors for wind farm," UPI, April 12, 2012, at http://www.upi.com/Business_News/Energy-Resources/2012/04/12/Cape-Wind-picks-contractors-for-wind-farm/UPI-87991334261408/.

[86] Andrew Meggison, "Wind Energy to Cost More than Fossil Fuels in Massachusetts," Gas2.org, April 5, 2012, at http://gas2.org/2012/04/05/wind-energy-to-cost-more-than-fossil-fuels-in-massachusetts/.

His failed venture has to be described as one of the nation's most expensive alternative energy boondoggles ever.

In May 2008, Pickens announced that his oil company, Mesa Power LP, would order 687 wind turbines, or 1,000 megawatts of capacity, from GE, at a cost of about $2 billion, a decision that a *New York Times* editorial at the time suggested President George W. Bush should carefully heed for policy purposes.[87] When Pickens went public with his plans, he boasted that by 2012, he would be able to expand the wind farm in west Texas to a gigantic 4,000 megawatts, about four times the output of a typical nuclear power plant.

At the height of his enthusiasm for wind turbine power, Pickens created a website to promote his "Pickens Plan" solution for a US energy policy that was intended to wean the US off foreign oil.[88] In total, he spent some $58 million to broadcast a series of television commercials promoting his agenda, and he appeared all over cable television news to promote his idea that wind power was a renewable energy that could save America from energy dependence on foreign oil.

A key Pickens television commercial began by tracking historically that the US imported 24 percent of all oil consumed in the country, growing to 42 percent in 1970 and "almost seventy percent" today and "climbing every minute." Pickens' some-

---

[87]   Editorial, "T. Boone Pickens Rides the Wind," New York Times, July 22, 2008, at http://www.nytimes.com/2008/07/22/opinion/22tue3.html?_r=1&ref=tboonepickens.

[88]   "Pickens Plan," PickensPlan.com, at http://www.pickensplan.com/media/?bcpid=1640183817&bclid=1641831862&bctid=1651750502.

what inflated numbers also assert that "over $700 billion leaves this country to foreign nations every year," an amount the commercial argues is "four times the cost of the Iraqi War."

The "Pickens Plan" in its heyday was also enthusiastic about converting eighteen-wheeler commercial trucks to be driven on natural gas, with a goal of converting 300,000 of the nation's fleet of 6.5 million long-haul trucks to run on natural gas. He was one of the first prominent oilmen to claim that electric batteries will be the ultimate solution for automobiles. Pickens openly acknowledged the financial difficulties: billions of dollars would have to be spent to modernize electric grids throughout the country, modify long-haul trucks, and provide a natural gas infrastructure of service stations around the country to create a new generation of liquid gas-driven, battery-powered cars.

The "pillars" of the Pickens Plan listed on his website included:

- Create millions of new jobs by building out the capacity to generate up to twenty-two percent of our electricity from wind. And adding to that with additional solar capacity
- Building a twenty-first century backbone electrical grid
- Providing incentives for homeowners and the owners of commercial buildings to upgrade their insulation and other energy savings options

- Using America's natural gas to replace imported oil as a transportation fuel

Anyone who has driven California has seen hundreds of abandoned wind turbines, built since the 1970s as a result of various tax-incentive subsidies that have attempted to promote the alternative energy or renewable energy agendas of past decades. Despite this, Pickens pleaded, "I've been an oilman all my life. But this is one emergency we can't drill our way out of."[89]

Unfortunately, Pickens failed to convince the federal government or the state of Texas to spend the hundreds of millions and possibly billions of dollars needed to connect the Pickens-built wind farm to the electrical grid in Dallas. As a result, a lot of wind turbines were left blowing in the dusty Texas wind. In July 2009, Pickens officially abandoned plans to build in Pampa.[90] And in the final analysis, he had no choice but to face the sad prospect of taking approximately a $2 billion loss on his wind turbine adventure.[91] Pickens tried to cut his losses by negotiating with GE to cut his massive order for wind turbines by more than half.[92]

---

[89] "T. Boone Pickens takes to the skies," *The Economist*, op.cit.

[90] Kate Galbraith, Pickens Drops Plan for Largest Wind Farm," *New York Times*, "Green" Blog, July 7, 2009, at http://green.blogs.nytimes.com/2009/07/07/pickens-drops-plan-for-largest-wind-farm/.

[91] Associated Press, "Pickens orders $2 billion in wind turbines," MSNBC, May 15, 2008, at http://www.msnbc.msn.com/id/24654895.

[92] Elizabeth Souder, "T. Boone Pickens cuts order for wind turbines, puts Panhandle wind farm on hold," *Dallas Morning News*, Jan. 13, 2010, at http://www.dallasnews.com/news/state/headlines/20100112-T-Boone-Pickens-cuts-order-915.ece.

But Pickens was loathe to give up his wind turbine dreams entirely. In April 2012, he announced he was proceeding to build a 377-megawatt wind farm in Texas, three years after shelving plans for the Pampa project that would have been some ten times larger, had it succeeded. Pickens decided to go ahead after Wind Energy Transmission Texas LLC, a joint venture company, agreed to build a transmission line to carry the power from the Pickens wind farm to utility providers in the state.[93]

Deere & Co. Abandons Wind Energy Project

In 2010, Deere & Co. also abandoned a costly boondoggle in the wind energy business, providing more evidence that wind turbine energy is marginal at best in its commercial potential.

On Aug. 31, 2010, Deere announced its intention to sell its wind energy business to a subsidiary of Exelon for $900 million, as reported by the Associated Press.[94] Originally, the company saw the wind business "as an extension of its agricultural work, with projects located in rural areas," as the AP noted. Deere had invested over $1 billion in the

---

[93] Andrew Herndon, "Pickens Reviving Plans for Texas Wind Power at Smaller Scale," Bloomberg, April 4, 2012, at http://www.bloomberg.com/news/2012-04-04/pickens-reviving-plans-for-texas-wind-projects-at-smaller-scale.html.

[94] Associated Press, "Deere sells wind energy business for $900M," *State Journal-Register*, Springfield, IL, Aug. 31, 2010, at http://www.sj-r.com/breaking/x833264297/Deere-sells-wind-energy-business-for-900M.

wind energy project over the last five years, buying much of its wind turbine equipment from Suzlon Energy, a company in India.

According to the AP, the Deere wind turbine business includes the physical infrastructure needed to operate thirty-six completed plants in eight states, with an operational capacity of 735 megawatts, enough, according to Exelon estimates, to power nearly 184,000 homes. In selling the wind turbine business, Deere anticipated recording a $25 million loss in the fourth quarter of 2010. In abandoning the wind turbine business, Deere decided to concentrate on what it does best – making farm equipment.

At the time of the sale, Exelon, the largest operator in the United States, was just entering the wind turbine business, attempting to be a wholesale marketer of wind energy in Illinois, Pennsylvania and West Virginia.

EPA Doubles the Amount of Ethanol Allowed in Gasoline

On April 2, 2012, the Environmental Protection Agency gave approval for cars and light trucks manufactured in 2007 and later to begin using 15 percent ethanol, known as E15, in a decision that pushed up the price of corn dramatically.

The EPA decision allowed a 50 percent increase from the current permitted limit of 10 percent ethanol in gasoline. According to the EPA statement, a decision on the use of E15 for cars and light trucks manufactured between 2001 and 2006 will be made after additional testing is completed in Nov. of 2012.

"Our nation needs E15 to reduce our dependence on foreign oil – it will keep gas prices down at the pump and help to end the extreme fluctuations in gas prices caused by our reliance on fuel from unstable parts of the world," proclaimed Tom Buis, the chief executive officer of Growth Energy, an ethanol trade group.[95]

The EPA decision regarding E15 was made in response to a request in March 2009 by Growth Energy, a coalition of US ethanol supporters, and fifty-four ethanol manufacturers who had applied for a waiver to increase the allowable amount of ethanol in gasoline from E10 to E15.[96]

Poor Die in Africa Because US Produces Ethanol

The poor are dying of famine in Third World countries such as Africa because of an Obama administration political agenda to produce ethanol as a renewable fuel substitute for gasoline.

---

[95] Associated Press, "EPA allows ethanol makers to register E15, moving closer to approval of 15 percent ethanol gas," *Washington Post*, April 2, 2012, at http://www. washingtonpost.com/national/energy-environment/epa-allows-ethanol-makers-to-register-e15-moving-closer-to-approval-of-15-percent-ethanol-gas/2012/04/02/gIQAO3rVrS_story.html.

[96] Environmental Protection Agency, "E15 (a blend of gasoline and ethanol," EPA.gov, last updated on April 2, 2012, at http://www.epa.gov/otaq/regs/fuels/additive/e15/.

The Obama administration's mandates for the use of ethanol are "immoral," Robert Bryce, a writer on ethanol for *Energy Tribune*, told the author in an email written in April 2009, following an article Bryce had written charging that ethanol drives food prices higher.[97] "We are burning food to make motor fuel at a time when there's a growing global shortage of food and no shortage of motor fuel," Bryce said. "The corn ethanol scam is not an energy program. It is a massive farm subsidy program masquerading as an energy program."

A controversial 2009 report released by the Congressional Budget Office, or CBO, documented that the increasing demand for corn to produce ethanol contributed between 10 to 15 percent for an overall 5.1 percent increase in the price of food from April 2007 to April 2008, as measured by the Consumer Price Index.[98] "Producing ethanol for use in motor fuels increases the demand for corn, which ultimately raises the prices that consumers pay for a wide variety of foods at grocery stores, ranging from corn syrup sweeteners found in soft drinks to meat, dairy, and poultry products," the CBO concluded.

An International Monetary Fund (IMF) assessment was even more pessimistic. "With respect to food, biofuels policies in some advanced economies are spilling over to the price of

---

[97] Robert Bryce, "Aventine Goes Down the Drain, Another Study Finds Ethanol Drives Food Prices Higher," EnergyTribune.com, April 10, 2009, at http://www. energytribune.com/articles.cfm?aid=1575.

[98] Congressional Budget Office, "The Impact of Ethanol Use on Food Prices and Greenhouse-Gas Emissions," April 2009, at http://www.cbo.gov/sites/default/files/ cbofiles/ftpdocs/100xx/doc10057/04-08-ethanol.pdf.

key food items, particularly corn and soybeans," John Lipsky, First Managing Director of the IMF, told the Council on Foreign Relations, on May 8, 2008. "IMF estimates suggest increased demand for biofuels accounts for 70 percent of the increase in corn prices and 40 percent of the increase in soybean prices."[99]

In an article entitled "How Biofuels Could Starve the Poor," published in the Council on Foreign Relations' *Foreign Affairs* magazine in its May/June 2007 issue, economists C. Ford Runge and Benjamin Senauer concluded that if the prices of staple foods increase because of the demand for biofuels, "the number of food-insecure people in the world would rise by over sixteen million for every percentage point in the real prices of staple foods." Runge and Senauer projected that as many as 1.2 billion people could be chronically hungry by 2025, 600 million more than previously projected, with the increase being due to the production of biofuels.[100]

Ethanol Producers Go Broke

Despite heavy government subsidies, the history of the ethanol industry in the United States is that even major producers cannot make a profit.

---

[99] Remarks by John Lipsky, First Deputy Managing Director, International Monetary Fund, at the Council on Foreign Relations, New York City, May 8, 2008, "Commodity Prices and Global Inflation," International Monetary Fund, IMF.org, at http://www.imf.org/external/np/speeches/2008/050808.htm.

[100] C. Ford Runge and Benjamin Senauer, "How Biofuels Could Starve the Poor," *Foreign Affairs* magazine, May/June 2007, at http://www.foreignaffairs.com/articles/62609/c-ford-runge-and-benjamin-senauer/how-biofuels-could-starve-the-poor.

The *Houston Chronicle* reported in May 2009 that White Energy, the largest ethanol producer in Texas, had filed for a Chapter 11 bankruptcy.[101] The White Energy bankruptcy adds to a string of ethanol industry bankruptcies that have called into question the economic viability of biofuel, despite federal government subsidies that amount to forty-five cents per gallon and a federal mandate that required US gasoline producers to use twelve billion gallons of ethanol in 2009, with the requirement increasing to fifteen billion gallons by 2015.

A Congressional Budget Office report issued in April 2009 concluded the "break-even ratio" of the price per gallon of retail gasoline to the price per bushel of corn is currently about 0.9. In other words, unless the price of gasoline is more than 90 percent of the price of a bushel of corn, it is not profitable to produce ethanol. When corn trades at about $5.78 a bushel, gasoline would have to cost about $5.20 a gallon for the production of ethanol to be profitable, even with government subsidies.

Finally, on Jan. 3, 2012, Congress adjourned without extending the multi-billion dollar subsidy for ethanol, thus allowing an ethanol subsidy that had been in place for more than thirty years to expire. In those thirty years, the ethanol subsidies

---

[101] Brett Clanton, "Ethanol bankruptcy filing a blow to biofuels industry," *Houston Chronicle*, May 11, 2009, at http://www.chron.com/business/energy/article/Ethanol-bankruptcy-filing-a-blow-to-biofuels-1530431.php.

totaled over $20 billion.[102] Yet, thirty years after Congress began subsidizing ethanol, no viable commercial ethanol energy has emerged in the United States. Even with legislative demands to include ethanol in gasoline fuel and billions of dollars in ethanol subsidies, scores of ethanol companies have gone bankrupt.

The truth is that biofuels are not necessarily energy efficient. The production of ethanol may burn up more hydrocarbon fuel than it saves. Consider the different uses of hydrocarbon fuels needed to convert corn into ethanol. Corn has to be planted, grown and harvested. Then corn needs to be transported to an ethanol plant and converted to ethanol through a chemical process that relies on hydrocarbon fuels.

An analysis conducted by David Pimentel, a professor of ecology and agriculture at Cornell University, and Tad Patzek, a professor of civil and environmental engineering at the University of California, Berkeley,[103] took into account the production of pesticides and fertilizers needed for the growing of crops, the running of farm machinery and irrigation, the grounding and transporting of crops, and the fermenting and distilling of ethanol from the water mix. Their conclusions were that corn requires 29 percent more hydrocarbon energy than the fuel produced, switch

---

[102] National Public Radio, "Congress Ends Era of Ethanol Subsidies," Jan. 3, 2012, at http://www.npr.org/2012/01/03/144605485/congress-ends-era-of-ethanol-subsidies.

[103] Reported in: Susan S. Lang, "Cornell ecologist's study finds that producing ethanol and biodiesel from corn and other crops is not worth the energy," Cornell University News Service, July 5, 2005, at http://www.news.cornell.edu/stories/July05/ethanol.toocostly.ssl.html.

grass requires 45 percent more, and wood biomass requires 57 percent more. The same goes for soybean plants used to produce biodiesel fuel (27 percent more hydrocarbon fuel used than produced) and sunflower plants (118 percent more hydrocarbon fuel used). The analysis did not factor in the additional costs in federal and state subsidies that are passed on to consumers in the form of taxes required to pay for the subsidies.

The Solyndra Bankruptcy – An Obama Administration Energy Scandal

When President Obama touts the "green economy," the mainstream media bends over backwards to give him extensive coverage. But when "green economy" ventures go bust, the mainstream media buries the story.

A prominent example was Solyndra, Inc, a maker of solar panels, headquartered in Fremont, California. In 2009, Solyndra received $535 million in a Department of Energy loan guarantee, in a ceremony attended by Vice President Joe Biden, Energy Secretary Steven Chu, and California Governor Arnold Schwarzenegger. Then, on May 26, 2010, President Obama personally toured the plant and California Senator Barbara Boxer proclaimed Solyndra to be the future not only of California, but also of the US economy.

Unfortunately, on Aug. 31, 2011, Solyndra shut the doors to its California headquarters and declared bankruptcy.[104]

---

[104] Matthew Mosk and Ronnie Greene, "White House-Backed Solar Energy Company Collapses," ABC News, Aug. 31, 2011, at http://abcnews.go.com/Blotter/white-house-backed-solar-solyndra-company-collapses/story?id=14420755#.T4sGKu05eS1.

Solyndra claimed the reason for the bankruptcy was cheaper foreign competition from China in manufacturing solar panels. But industry experts had a different explanation. Axiom Capital Management's solar power analyst, Gordon Johnson, told Bloomberg that the supply of photovoltaic panels is expected to climb to almost triple the level of demand in 2011, crashing prices in the industry.[105] "It could be Armageddon," Johnson said. "Demand is about to fall at a time when you're going to have a significant increase in supply. In a commoditized industry, that is a formula for disaster."

What emerged from the Solyndra bankruptcy was a pattern of the Obama administration giving massive financial benefits to Obama campaign contributors who were willing to venture into green energy businesses. An April 2012 Treasury Department report, entitled "Consultation on Solyndra Loan Guarantee Was Rushed," revealed that the Department of Energy cut out Treasury officials from raising financial issues involving Solyndra, ignoring advice from Treasury officials and limiting the Treasury Department's opportunity to review the high-priced, high-financing-risk of what critics called "an Obama pipe dream."[106] A top Obama fundraiser, George Kaiser of Tulsa, Oklahoma, had bankrolled Solyndra, leading to charges that corruption,

---

[105] Ben Sills, "Solar Panel Makers Face Supply-Glut," Bloomberg, Nov. 15, 2010, at http://www.bloomberg.com/news/2010-11-16/solar-panel-makers-face-supply-glut-armageddon-chart-of-day.html.

[106] Jim Kouri, "Obama administration corruption in Solyndra deal confirmed," Examiner.com, April 10, 2012, at http://www.examiner.com/law-enforcement-in-national/obama-administration-corruption-solyndra-deal-confirmed.

not jobs, was the motivation behind the Solyndra government financing from the beginning.

On Sept. 8, 2011, just two days after the company formally declared Chapter 11 bankruptcy, the FBI and the Energy Department's inspector general's office executed a search warrant at the Fremont, California, headquarters of solar panel maker Solyndra, seizing the company's records and computers.[107] In Dec. 2011, the *Wall Street Journal* reported that over the past several months, at least seven solar panel manufacturers, in additional to Solyndra, had filed for bankruptcy or declared insolvency.[108] The problem was predictable – Chinese low-cost manufacturers had entered the solar panel business, undercutting US and EU manufacturers pursuing the clean energy dreams of the Obama administration and their leftist political supporters.

Top Leftist Altruist Goes Bust on Green Energy Investments

In Dec. 2009, David Gelbaum, a major donor to the Sierra Club and the American Civil Liberties Union, as well as several military assistance foundations with leftist political bents, announced that after donating $389 million to these groups between 2005

---

[107] "FBI, Energy Department raid offices of solar-panel maker Solyndra," *Los Angeles Times*, Sept. 8, 2011, at http://latimesblogs.latimes.com/money_co/2011/09/joint-fbi-energy-department-raid-on-solyndra-offices.html.

[108] Yuliya Chernova, "Dark Times Fall on Solar Sector," *Wall Street Journal*, Dec. 27, 2011, at http://online.wsj.com/article/SB10001424052970204552304577117140511996840.html.

and 2009, he had to cut back because his investments in alternative-energy firms "laced me in a highly liquid position," according to a *Wall Street Journal* report.[109] He made his fortune as a mathematician working in a Wall Street hedge fund, but now his commitment to renewable energy had cost him dearly. Gelbaum, also a major donor to the Democratic National Party, indicated to the *Wall Street Journal* that the Quercus Trust, an investment fund he runs, was down almost 57 percent over an eighteen-month period from 2008 to 2009.

In Nov. 2008, GreenTechMedia.com identified thirty-four green technology companies that had received Quercus Trust funding.[110] "Entrepreneurs who have received money say Gelbaum takes a long term, holistic view of the market and is patient enough to put money into an investment that might pay off well beyond five years," GreenTechMedia.com wrote. "He [Gelbaum] is also not seeking attention." In Jan. 2009, GreenTechMedia identified that Quercus Trust investments had been placed in forty-seven green technology companies. "Entrepreneurs who have received money from the trust say Gelbaum is not investing in these companies as a way to evangelize green or as a form of charity," GreenTechMedia.com wrote.[111]

---

[109] Siobhan Hughes, "ACLU, Sierra Club Donor to Cut Funding," *Wall Street Journal,* Dec. 10, 2009, at http://online.wsj.com/article/SB126039786736984375.html.

[110] Michael Kanellos, "The Secret Life of the Quercus Trust," GreenTechMedia.com, Nov. 7, 2008, at http://www.greentechmedia.com/articles/read/the-secret-life-of-the-quercus-trust-5135/.

[111] Michael Kanellos, "Quercus Trust Continues to Grow: 47 Found in Portfolio," GreenTechMedia.com, Jan. 5, 2009, at http://www.greentechmedia.com/articles/read/quercus-trust-continues-to-grow-47-found-in-portfolio-5453/.

That Gelbaum's fortune was diminished by his ideological enthusiasm for green energy companies was made clear by a *USA Today* article published in Nov. 2010, under the title "Donor's millions for military causes drying up."[112] The newspaper noted that Gelbaum had donated $450 million to environmental causes for several years and invested $500 million in clean technology. In 2005, he funded charities that assisted troops, veterans, and their families through Sierra Club and an organization Gelbaum created, called the Iraq and Afghanistan Deployment Impact Fund. Gelbaum told *USA Today* that his personal funds for charity had dried up because he had lost or remained at risk for hundreds of millions of dollars he invested in green-technology start-up companies that had done poorly or have not yet become commercially viable.

NIMBY Syndrome Blocks Wind Turbine and Solar Panel Farms

Ever since former Massachusetts Senator Ted Kennedy objected to putting wind turbines off his beloved Cape Cod, the NIMBY, or "Not In My Back Yard," syndrome has been a major obstacle to the expansion of wind and solar power around the world. The truth is that wind and solar power require a massive amount of space in order to generate the quantities of electricity needed to provide more than minimal energy.

---

[112] Gregg Zoroya, "Donor's millions for military causes drying up," *USA Today*, Nov. 22, 2010, at http://www.usatoday.com/news/military/2010-11-22-militarycharities22_ST_N.htm.

In an article entitled "Renewable Energy, Meet the New Nimbys," reporter Jeffrey Ball wrote in the *Wall Street Journal* that, "Even as Americans tell pollsters they are eager for alternatives to fossil fuel, some are fighting proposals for solar and wind projects and for the thousands of miles of transmission lines that would be needed to carry the cleaner energy to market."[113] The new backlash, Ball noted, was fueled by worries that renewable-energy projects would need to occupy vast amounts of land to produce significant amounts of power. He reported that California, considering a proposal to produce a third of its electricity from renewable sources by 2020, would have to build vast solar energy plants in the Mojave Dessert. As enthusiastic as Californians might be to get solar energy, environmentalists resist the massive intrusion the solar plants would impose, even on a region as remote as the Mojave Dessert.

National Geographic examined how large an area for windmill technology would be needed if New York City were to abandon coal and natural gas to generate 60 percent of the city's needs for electricity. The answer was 10.6 square miles, an area larger than southern Manhattan from the tip of the island through Greenwich Village, which would have to hold some 6,800 turbine windmills, each capable of generating 1.5 megawatts of electricity.[114] Yet, to deliver the

---

[113] Jeffrey Ball, "Renewable Energy, Meet the New Nimbys," *Wall Street Journal*, Sept. 4, 2009, at http://online.wsj.com/article/SB125201834987684787.html.

[114] Michael Parflit, "Future Power: Where Will the World Get Its Next Energy Fix?" *National Geographic*, no date, at http://ngm.nationalgeographic.com/ngm/0508/

same amount of electricity with solar power would take an area of seventy-four square miles, an area stretching from about 59th Street north to the tip of the island in a square block that would reach across into New Jersey on the west and include the Bronx on the east. The installation would involve over 145 million solar panels, each delivering 175 watts of power. By comparison, that quantity of electricity would take four nuclear reactors capable of delivering 1,000 megawatts each, with each plant taking up about two square miles.

Still, when the wind does not blow, windmill technology is no more effective than a sailboat caught in the doldrums. A modern giant windmill, standing about 150 meters high (about 500 feet) with a blade diameter of about 100 meters (328 feet, slightly more than one football field) can generate about two megawatts of electricity when the wind blows hard (about ten miles per second). But when the wind blows moderately, say at one mile per second, the windmill hardly produces any electricity at all, not even enough to power an average dishwasher. Experts estimate that it would require 1,500 giant windmills operating with the wind always blowing at full capacity to produce as much energy as one nuclear reactor of 1,500 megawatts with a reactor that would be at most only a few meters high and wide.[115]

---

feature1/index.html.

[115] "Calculation of the power produced by a windmill," Ecolo.org, no date, at http://www.ecolo.org/documents/documents_in_english/WindmillFormula.htm.

A 2007 study titled "Calculating the Real Cost of Industrial Wind Power,"[116] produced in Bruce County, Ontario, Canada, examined data from wind power generated on an industrial basis in Europe over the last ten years. The study concluded, "As the public increasingly learns the real costs of wind turbine development, publically subsidized industrial wind projects are rapidly becoming unacceptable."

The study noted that in Denmark, which has one of the world's highest concentrations of wind turbines, approximately 80 percent of the wind energy produced has to be sold to Denmark's neighbors, Norway and Sweden, "at a price far below the cost of production in order to stabilize the grid because it is produced during periods of low consumer demand." Conversely, the study observed, Denmark is frequently forced to buy hydro and nuclear power from its neighbors. "The net outcome," the Ontario study concluded, "is that Denmark with the highest amount of installed wind energy has the highest consumer electricity charges in Europe. Danish households already pay 100 percent more for their electricity than other European customers."

Toward a Comprehensive Energy Policy: EPA Shuts Down Coal Plants?

The national energy policy implemented by the Obama administration has heavily favored green energy technologies,

---

[116] Keith Stelling, "Calculating the Real Cost of Industrial Wind Power: An Information Update for Ontario Electricity Customers," Friends of Arran Lake Wind Action Group, Bruce County, Ontario, Canada, November 2007, at http://www.wind-watch.org/documents/calculating-the-real-cost-of-industrial-wind-power/.

despite the failure of renewable energies such as wind turbine and solar power. What the nation's thirty-year experience with ethanol has demonstrated is that government regulations and subsidies are not sufficient to make problematic energy technologies into commercially viable realities. Scandals such as Solyndra reinforce the point, demonstrating that corruption, not sound energy policy, is the most likely result when ideology, not practical energy realities, dictates the nation's energy policy.

Through 2013, the EPA plans to implement new rules designed to curb pollution from coal-fired power plants. Experts estimate that the regulations will cost utilities up to $129 billion and force the retirement of up to 20 percent of the nation's coal capacity. Given that coal now powers approximately 45 percent of US electric power, the new EPA regulations inevitably mean the closing of possibly dozens of electric plants, and higher electric bills.[117] Again, the EPA plan indicates that the Obama administration places ideology above economic efficiency when dictating the nation's energy policy.

Championing oil produced from Canadian tar sands or US oil shale is the type of technological innovation unanticipated when Shell Oil's Hubbert first drew his peak production graph. But when the Obama administration decided to block the Keystone XL pipeline from bringing that oil from

---

[117] Brad Plumer, "Getting Ready for a wave of coal-plant shutdowns," *Washington Post*, Aug. 19, 2011, at http://www.washingtonpost.com/blogs/ezra-klein/post/getting-ready-for-a-wave-of-coal-plant-shutdowns/2011/08/19/gIQAzkZ0PJ_blog.html.

Canada to Texas, US energy policy returned to being driven by the politics of the environmental movement. At its core, the Obama administration energy policy displays a hostility to hydrocarbon fuels that has already cost the US taxpayer countless billions in wasted loan guarantees, pointless subsidies, and political corruption.

6

"Drill, Baby, Drill" – Achieving USA Energy Independence Now

President Barack Obama has repeatedly claimed the United States consumes more than 20 percent of the world's oil reserves, but we have less than 2 percent of all the world's oil reserves.

Clearly, we must ask, "How can Obama possibly know exactly how large the US or the world oil reserves truly are?" This is a common mistake made by peak production theorists. Implicit in the concept that we are running out of hydrocarbon fuels is the assumption that we know the quantity of hydrocarbon fuels that exist worldwide. Moreover, the assumption is that those hydrocarbon resources are finite, not subject bo being expanded by technological advances. Otherwise, how could we know we were running out?

In evaluating Obama's claim, the *Washington Post* pointed out that two arms of the Interior Department – the US Geological Survey (for onshore estimates) and the Bureau of Ocean Energy Management (for offshore estimates), reported the US in 2011 had 219 billion barrels of "undiscovered technically recoverable resources" that may be recovered depending in part on technology and/or the price of oil. This is ten times more than the twenty-one billion barrels of oil the Energy Information Administration claims the United States had in 2011 when counting "proven reserves." In other words, as the *Washington Post* pointed out, "These estimates change over time."

A greater example of this is the Bakken Formation that stretches across three states in the northern United States and into southern Canada.[118]

The Bakken Formation

The Bakken Formation, discovered in the 1980s and 1990s, was initially thought to have only a limited amount of oil, scattered between layers of shale and sandstone. The US Geological Survey estimated in 1995 that the Bakken Formation had only about 151 million barrels of recoverable oil.

---

[118]   Glenn Kessler, "US oil resources: President Obama's 'non sequitur facts,'" *Washington Post*, March 15, 2012, at http://www.washingtonpost.com/blogs/fact-checker/post/us-oil-resources-president-obamas-non-sequitur-facts/2012/03/14/gIQApP14CS_blog.html.

Then, with advances in drilling technology, the US Geological Survey reassessed the quantity of recoverable oil in the Bakken Formation. A USGS assessment released in April 2008 concluded that the Bakken Formation may have an estimated 3 to 4.3 billion barrels of technically recoverable oil, a 2,800 percent, or twenty-eight-times increase in the amount of oil recoverable compared to the agency's initial 1995 assessment. The Energy Information Administration has officially attributed the success of horizontal drilling and fracturing efforts in Montana as the reason a decision was made to reevaluate the 1995 assessment.[119]

"Oil production from shale plays, particularly in the Bakken shale in North Dakota, has been rising rapidly," Richard Newell, the EIA administrator, told the House Committee on Natural Resources, on March 17, 2011.[120] "Using horizontal drilling and hydraulic fracturing, operators increased Bakken production from about 3,000 barrels per day in 2005 to 137,000 barrels per day in 2009 and 225,000 barrels per day in 2010." Newell told Congress that the government currently estimates there are nearly twenty-four billion barrels of technically recov-

---

[119] "Technology-Based Oil and Natural Gas Plays: Shale Shock! Could There Be Billions in the Bakken?" US Department of Energy, Energy Information Administration, Office of Oil and Gas, Reserves and Production Division, November 2006.

[120] Statement of Richard Newell, Administrator, Energy Information Administration, US Department of Energy, before the Committee on Natural Resources, US House of Representatives, March 17, 2011, at http://www.eia.gov/neic/speeches/newell_03172011.pdf.

erable crude oil in Bakken and three other producing shale oil formations in the United States.

"The domestic grade oil and natural gas industry has undergone a technological revolution that has revitalized the resource base in the onshore lower-forty-eight states," Newell continued. "The use of horizontal drilling in conjunction with hydraulic fracturing has greatly expanded the ability of producers to profitably produce crude oil and natural gas from low permeability geologic formations, particularly shale oil formations." As a result of this technological revolution, US natural gas reserves grew 63 percent between 2000 and 2010, increasing from 167.4 trillion cubic feet at the start of 2000 to 272.5 trillion cubic feet at the start of 2010, the highest since 1971.

"The Bakken Formation estimate is larger than all other current USGS oil assessments of the lower 48 states and is the largest 'continuous' oil accumulation ever assessed by the USGS," said the USGS press release making the April 2008 announcement.[121] The Bakken Formation lies in "Williston Basin," a geological formation in the north central United States, underlying much of North Dakota, eastern Montana, northwestern South Dakota, and southern Saskatchewan and Manitoba, Canada.

---

[121] US Department of the Interior, US Geological Survey, "3 to 4.3 Billion Barrels of Technically Recoverable Oil Assessed in North Dakota and Montana's Bakken Formation – 25 Times More than 1995 Estimate," April 10, 2008, at http://www.usgs.gov/newsroom/article.asp?ID=1911

US - The Saudi Arabia of Natural Gas

In 2008, Aubrey McClendon, the chief executive of major natural gas producer Chesapeake Energy Corporation, proclaimed, "Shale gas makes the United States the Saudi Arabia of natural gas."[122]

Another unanticipated hydrocarbon fuel resource when M. King Hubbard predicted in 1956 that peak production rates were soon to be reached, newly explored shale oil and shale gas reserves in North America promise to provide abundant domestic reserves adequate to meet US energy needs for hundreds of years to come. Yet, in an era where the Obama administration is pressing a carbon-hysteria agenda masking an ideological preference for green renewable energy alternatives such as wind and solar power, President Obama is unlikely to tout abundant shale oil reserves as a solution to provide cheap energy for decades to come.

In sharp contrast, the green industries – including ethanol and other biofuels, as well as wind turbines and solar power – largely collapsed in 2011, amidst scandals that tied billions of dollars in public funding and tax breaks to Democratic donors and fundraisers for President Obama. To the dismay of the Obama administration and their friends, 2011 saw a boom, not in ethanol, not in biofuels, not in wind or solar energy, but in the

---

[122] ICIS News, "Shale gas can meet US needs for 100 years – study," ICIS.com, July 30, 2008, at http://www.icis.com/Articles/2008/07/30/9144315/shale-gas-can-meet-us-needs-for-100-years-study.html.

oil and gas arena, where technological advances have truly created a horizon for sustainable profits into the foreseeable future, barring political intervention and disruption from an ideologically driven EPA.

In Dec. 2011, the *Wall Street Journal* reported that the boom in low-cost natural gas obtained from shale is driving investment in plants that use gas for fuel or as a raw material, setting off a race by states to attract such factories and the jobs they create.[123] The article noted that shale gas now accounts for more than one-third of all US natural-gas production; the surge in production has pushed down US natural-gas prices, from a high of about $15 per million British thermal units six years ago, to today when near-term futures prices have fallen below $3.20. Whether the green ideologes in the Obama administration like it or not, the energy play for 2012 and beyond remains in the oil and natural gas fields, not in green energy technologies.

The Age of Small Nuclear Reactors Beginning to Dawn

Bill Gates of Microsoft Fame has financed TerraPower LLC, a company created to build small-scale nuclear reactors that theoretically could power a local community for decades at a time without having to be refueled. Gates is intrigued by the poten-

---

[123]   James R. Hagerty, "Shale-Gas Boom Spurs Race," *Wall Street Journal*, Dec. 27, 2011, at http://online.wsj.com/article/SB1000142405297020484450457710042125300051 22.html?mod=ITP_pageone_1.

tial for small nuclear reactors to produce cheap, zero-carbon energy and its ability to turn what is a waste product (depleted uranium) into fuel.

The TerraPower traveling-wave reactor is designed to be buried in the ground, where it would run for 100 years. Describing how the reactor would work, the *Wall Street Journal* explained that enriched uranium would shoot neutrons into the depleted uranium making up approximately 90 percent of the fuel. That process would produce plutonium, designed to burn slowly in a controlled reaction that would continue over many years without the need of human intervention. The *Wall Street Journal* also pointed out that large supplies of depleted uranium are available as a byproduct of today's water-cooled nuclear reactors.[124]

Another pioneer in the small nukes business, Hyperion Power Generation, Inc., was formed to market a small, modular, non-weapons grade nuclear power generator created by Dr. Otis "Pete" Peterson at the Los Alamos National Office in New Mexico, with the goal of powering industrial plants, military bases, hospitals, government complexes, and college campuses. The Hyperion website touts a small reactor – 1.5 meters in length and width, 2.5 meters in height – that produces enough electricity to power 20,000 average American homes. The module can be buried underground, "out of sight and harm's way," and is equally capable of being transported by train, ship or

---

[124] Robert A. Guth, "A Window Into the Nuclear Future," *Wall Street Journal*, Feb. 27, 2011, at http://online.wsj.com/article/SB100014240527487044090004576146061231899264.html?mod=WSJ_hp_LEFTTopStories.

truck. Buried underground, the Hyperion nuclear generator is designed to provide power for 7 to 10 years with minimal maintenance and no emission of so-called greenhouse gases.

The success of companies producing small nuclear reactors has led to the designation of a new industry grouping under the heading of "Small and Modular Nuclear Power Reactors," or SMRs. Small Modular Reactors have the advantage of providing power away from the large grids generating electricity throughout most of the United States today, according to the World Nuclear Association.[125]

---

[125] World Nuclear Association, "Small Nuclear Power Reactors," March 22, 2012, at http://www.world-nuclear.org/info/inf33.html.

Conclusion: The USA #1 in Oil Production?

On Sept. 11, 2011, Goldman Sachs issued a report predicting the United States would be the world's largest oil producing country by 2017. The Goldman Sachs report forecast that US daily production of oil would grow from a current 8.3 million barrels of oil per day to 10.9 million barrels by 2017, a level of production that would surpass both Saudi Arabia and Russia.[126] The report was a shock to peak oil believers in the oil industry, who had been conditioned to expect that the United States would be close to total oil depletion by 2017, not that the United States

---

[126] Keith Schaefer, "Goldman Sachs predicts that US will be world's largest producer of oil in 2017," Oil and Gas Investments Bulletin, Sept. 15, 2011, at http://oilandgas-investments.com/2011/top-stories/goldman-sachs-predicts-that-u-s-will-be-worlds-largest-producer-of-oil-in-2017/.

could possibly be the world's largest producer of oil within this decade.

Little noticed, data from the Energy Information Administration has documented that US reliance on foreign oil has actually shrunk in recent years, from over 60 percent in 2006 to under 50 percent in 2010.[127] The prolonged economic downturn, continuing since 2008, has reduced demand for oil in the US economy. However, Goldman Sachs concluded the effect of a slow economy was insufficient to explain the entire shift. US hydrocarbon liquids production, including both crude oil and liquid natural gas, has jumped roughly one million barrels per day between 2008 and 2011. Much of that has come from increased production in the onshore lower forty-eight states and reflects the significant contributions of America's independent producers. Independents currently produce 95 percent of the oil and gas wells in the United States. The investment bank report also noted that net import for natural gas was at its lowest point in seventeen years, at 10.8 percent, down from a peak of 16.4 percent in 2005.

On Aug. 28, 2008, WND columnist Eric Rush published a piece in which he countered the far left and Congressional Democrats who oppose exploration for new oil resources and have done so for decades by declaring, "Drill, Baby, Drill!"[128] We

---

[127] Energy Information Administration, "US Imports of Crude Oil and Petroleum Products," at http://www.eia.gov/dnav/pet/hist/LeafHandler. ashx?n=PET&s=MTTIMUS1&f=M.

[128] Eric Rush, "Drill, Baby Drill!" WND, Aug. 28, 2008, at http://www.wnd.

conclude here by echoing Eric Rush's call to action. It is time to end the cover-up and misinformation that has prevented the American public from knowing the truth about oil – that hydrocarbon fuels are abiotic in nature, produced by the earth naturally on a continuous basis, and that the quantity of abiotic hydrocarbons yet to be discovered suggests the world will never run out of oil or natural gas, exactly as Julian Simon predicted decades ago.

Consider this paragraph published by Simon in 1981:

> *Natural resources. Hold your hat—our supplies of natural resources are not finite in any economic sense. Nor does past experience give reason to expect natural resources to become more scarce. Rather, if the past is any guide, natural resources will progressively become less scarce and less costly, and will constitute a smaller proportion of our expenses in future years. And population growth is likely to have a long run beneficial* impact on the natural-resource situation.[129]

We must counter the Malthusians by learning how to think from the perspective of abundance, not scarcity. We should place our confidence in a private economy in which entrepreneurs and independent economic actors could adapt to market

---

com/2008/08/73559/.

[129] Julian L. Simon, *The Ultimate Resource* (Princeton, N.J.: Princeton University Press, 1981), "Introduction: What Are the Real Population and Resource Problems?" pp. 3-11, at p. 4.

conditions, seeking profit opportunities outside of government assistance. A market unrestricted by needless government regulations succeeds more than the highly regulated economy hostile to hydrocarbon fuels that the Obama administration seeks to impose upon us.

Air renews itself naturally, as does water. Why should oil or natural gas be different? The scenario of a modern industrial society, doomed to outlive the affordable hydrocarbon fuels that have made economic growth and prosperity possible, is consistent with a secular society desperate to replace God and Divine Providence with central planning imposed by a crushing state bureaucratic apparatus. In the 1950s, Sinclair Oil sold gasoline to motorists under a logo that featured a green dinosaur, while Shell Oil employed an executive who sought to prove the end of "fossil fuels" was at hand by adapting the graph of a normal distribution that is taught to every Statistics 101 college student. More than sixty years later, the world still has abundant hydrocarbon fuels, even though Sinclair Oil has dropped the dinosaur logo.

Major oil companies appear ready to drop fossil fuel illusions. A Shell Oil executive has expressed doubt on national television that peak production theory is correct. "The peak oil theory has really swamped the world," John Hofmeister, then the president of Shell Oil's US operations, said on CNBC's *Squawk Box* show on March 20, 2008. "God Bless Matt Simmons. His assumptions are correct based on his hypotheses, but his hypotheses are too narrow."[1] This is a remarkable admission from the company that produced M. King Hubbert. There is no reason America should be dependent upon foreign sources for oil. There is no reason we should be paying exorbitant amounts for a gallon of gasoline at the pump. Allowed to do their job without unnecessary government intervention, independent producers in conjunction with major oil companies should be able to provide Americans with an abundant supply of inexpensive energy for decades to come, especially if Americans are finally told the truth that oil and natural gas are not now and never were fossil fuels.

---

[1] "Shell Exec Says World Not Running Out of Oil," WND, March 20, 2008, at http://www.wnd.com/2008/03/59502/.

Auschwitz

A Doctor's Eyewitness Account

by Dr. Miklos Nyiszli
Translated by Tibere Kremer & Richard Seaver
Introduction by Bruno Bettelheim

When the Nazis invaded Hungary in 1944, they sent virtually the entire Jewish population to Auschwitz. A Jew and a medical doctor, Dr. Miklos Nyiszli was spared from death for a grimmer fate: to perform "scientific research" on his fellow inmates under the supervision of the infamous "Angel of Death," Dr. Josef Mengele. Nyiszli was named Mengele's personal research pathologist. Miraculously, he survived to give this terrifying and sobering account.

$14.95 Paperback

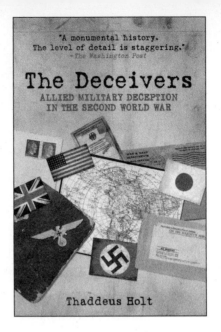

The Deceivers

Allied Military Deception in the Second World War

by Thaddeus Holt

Secret codes, ciphers, strategic misdirection, and more: Deception was one of the most powerful weapons utilized by the Allies in World War II. Here are all the amazing tricks and leaked misfortunes—many revealed for the first time—that helped lure the Axis powers into false, even dangerous, positions. The collection of incredible codes, surreptitious spies, and false battle plans is made all the more enjoyable by Thaddeus Holt's masterful writing, as well as the accompanying photos. His novel-like storytelling includes many illuminating profiles of the war's central figures and the roles they played in specific deceptive operations.

$22.95 Paperback

If the Allies Had Fallen

Sixty Alternate Scenarios of World War II

Edited by Dennis Showalter & Harold Deutsch
Introduction by Bill Forstchen

From the Munich crisis to the dropping of the first atom bomb, and from Hitler's declaration of war on the United States to the D-Day landings—historians suggest "what would have been" if key events in the war had gone differently. Written by an exceptional team of historians as if these world-changing events had really happened, realistic scenarios based on the true capabilities and circumstances of the opposing forces are projected with chilling implications. What if the Nazis had taken London? The alternate outcomes are fascinating and fully realized. *If the Allies Had Fallen* is a spirited and terrifying alternate history, and a telling insight into the dramatic possibilities of World War II. Includes detailed maps. Contributors include Thomas Barker, Harold Deutsch, Walter Dunn, D. Clayton James, Robert Love, Bernard Nalty, Richard Overy, Paul Schratz, Dennis Showalter, Gerhard Weinberg, Anne Wells, and Herman Wolk.

$26.95 Hardcover
$14.95 Paperback

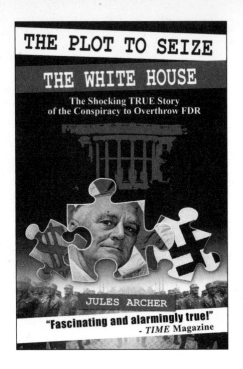

The Plot to Seize the White House

The Shocking TRUE Story of the Conspiracy to Overthrow FDR

by Jules Archer

Most people will be shocked to learn that in 1933 a cabal of wealthy industrialists—in league with groups like the K.K.K. and the American Liberty League—planned to overthrow the U.S. government in a fascist coup. Their plan was to turn discontented veterans into American "brown shirts," depose F.D.R., and stop the New Deal. They clandestinely asked Medal of Honor recipient and Marine Major General Smedley Darlington Butler to become the first American Caesar. He, though, was a true patriot and revealed the plot to journalists and to Congress. In a time when a President has invoked national security to circumvent constitutional checks and balances, this episode puts the spotlight on attacks upon our democracy and the individual courage needed to repel them.

$14.95 Paperback

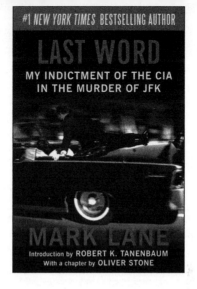

The Last Word

My Indictment of the CIA in the Murder of JFK

by Mark Lane
Introduction by Robert K. Tanenbaum
Contributions by Oliver Stone

Mark Lane tried the only U.S. court case in which the jurors concluded that the CIA plotted the murder of President Kennedy, but there was always a missing piece: How did the CIA control cops and secret service agents on the ground in Dealey Plaza? How did federal authorities prevent the House Select Committee on Assassinations from discovering the truth about the complicity of the CIA?

Now, *New York Times* best-selling author Mark Lane tells all in this explosive new book—with exclusive new interviews, sworn testimony, and meticulous new research (including interviews with Oliver Stone, Dallas Police deputy sheriffs, Robert K. Tanenbaum, and Abraham Bolden). Lane finds out firsthand exactly what went on the day JFK was assassinated. Lane includes sworn statements given to the Warren Commission by a police officer who confronted a man who he thought was the assassin. The officer testified that he drew his gun and pointed it at the suspect who showed Secret Service ID. Yet, the Secret Service later reported that there were no Secret Service agents on foot in Dealey Plaza.

The Last Word proves that the CIA, operating through a secret small group, prepared all credentials for Secret Service agents in Dallas for the two days that Kennedy was going to be there—conclusive evidence of the CIA's involvement in the assassination.

$24.95 Hardcover
$14.95 Paperback